T0339822

NIETZSCHE ON WAR

NIETZSCHE ON WAR

Rebekah Peery

Algora Publishing
New York

Library of Congress Cataloging-in-Publication Data —

Peery, Rebekah S.
  Nietzsche on war / Rebekah S. Peery.
     p. cm.
  Includes bibliographical references and index.
     ISBN 978-0-87586-711-3 (trade paper : alk. paper) — ISBN 978-0-87586-
712-0 (hard cover : alk. paper) — ISBN 978-0-87586-713-7 (ebook) 1. Nietzsche,
Friedrich Wilhelm, 1844-1900. 2. War (Philosophy) I. Title.
     B3318.W37P44 2009
     355.0201—dc22
                              2009009341

Front cover: Friedrich Wilhelm Nietzche, (1844-1900). © Bettmann/CORBIS

Printed in the United States

TABLE OF CONTENTS

INTRODUCTION

Given the increasing efforts in all ages to understand the nature of war — the possible causes, reasons, purposes — the reader of this book might question the possibility of any opportunity remaining for further enlightenment on the subject. The continuing impotence of men to comprehend the depth and breadth of this complex phenomenon, plus the rapid increase in the catastrophic effects of the power of war, make any and every new effort a moral imperative. Well into a new century, the subject is unavoidable. Repelled by the images and words — the anguish, the disbelief, the despair — we feel a compelling urgency to strengthen our efforts. The revulsion is becoming so intense and widespread that the overwhelming cruelty generated in war is serving to generate a response equally intense and widespread. Twenty-first century warfare is unacceptable and everything about it must be exposed.

More puzzling perhaps, to the reader, might be the question of "why Nietzsche?" Why appeal on the topic of war to this philosopher whose time preceded our own by more than a century? Nietzsche's thinking opened up new possibilities and opportunities in the arts, in the sciences, in history, in cultural studies, in philosophy, and in religion; thus he becomes the leading candidate — perhaps with challenging, rich, untapped, unexpected resources.

Nietzsche's style of thinking and writing is tantalizing but difficult. His scholarly background, plus the wide range of his interests, suggest that his qualifications are exceptional. In addition to his philosophical endeavors, he studied theology and religion, and insisted on the importance of rethinking philosophy "scientifically" and "historically." And, significantly for purposes of questioning him concerning war, he proclaimed that the new science of psychology was "now again the road to the basic problems." Nietzsche was never under any illusion that his ideas, and his treatment of those ideas, were anything but revolutionary and dangerous.

This book is based on the presupposition that had Nietzsche directly focused his critical powers on the urgent questions concerning hostile, violent war, his interpretations of the destructive, corrosive nature of such war would have matched his critique of Christianity in both substance and intensity.

From different perspectives the book offers new interpretations of Nietzsche's thinking — considering especially his ideas regarding power, values, nature, contrariety, language or words, truth and deception, religion, experience, sexuality and sexual politics — that could provide new and provocative approaches toward dealing with the rising menace of war. I have consulted selected earlier writers whom I believe could have influenced Nietzsche's own thinking on the subject of war. Also, I have considered a few later writers whose thinking appears relevant.

Particularly as philosopher, psychologist, philologist, and historian, Nietzsche provides the immediate and best access to his thoughts through his own words. (Citations from Nietzsche's works are referenced in the bibliography and cited in the text by initials in parentheses.)

My project has been to probe these "deep thoughts" and to piece together answers to my questions — keeping in mind that he himself did not directly and critically address the problem of hostile war.

Faced with the questions of war, there is little doubt that Nietzsche's primary concern would be to address the source, or origin, of such activities. He would agree with the necessity now to seek and reveal this origin — whatever the risks. As with the polluting of our air, which endangers breathing and destroys life, war is likewise polluting, which fouls, poisons, makes filthy the process of life. In one sense, Nietzsche's life and thinking were devoted to sounding the alarm to awaken humans to the forces rapidly developing to destroy nature, life, the body, culture.

We would not expect any ordinary response from him. Whether dreading or anticipating his words, typically we expect that these would be extraordinary. And there are multiple reasons for believing that he would give less consideration to so-called external events or situations and seek instead more understanding of internal, psychological experiences.

In speaking about war, words matter. If what is needed is a systematic analysis requiring many words, the Encyclopaedia Britannica probably serves best. A few paragraphs from both the Micropaedia and the Macropaedia offer a straightforward and sober introduction to the subject. First, from the Micropaedia:

> **War,** a state of usually open and declared armed hostile conflict between political units, such as states or nations or between rival political factions of the same state or nation. War is characterized by intentional violence on the part of large bodies of individuals who are expressly organized and trained to participate in such violence. Wars between nation-states may be fought to gain reparation for a particular injury; to acquire a particular territory or advantage; to gain recognition of a particular claim; or to achieve the extermination or unconditional surrender of the enemy. Wars of considerable duration have usually been divided into "campaigns" conducted in one area under one command for one season. These, in turn, have been composed of "battles," in which opposing forces come into direct contact for one or two days at a time (though battles have tended to last longer in modern warfare). With the progress of science and technology and consequent increases in the destructive power of armaments, war has had an increasingly catastrophic effect on human existence.

Second, from the Macropaedia:

> War in the popular sense is a conflict among political groups involving hostilities of considerable duration and magnitude. In

the usage of social science, certain qualifications are added. Sociologists usually apply the term to such conflicts only if they are initiated and conducted in accordance with socially recognized forms. They treat war as an institution recognized in custom or in law. Military writers usually confine the term to hostilities in which the contending groups are sufficiently equal in power to render the outcome uncertain for a time. Armed conflicts of powerful states with primitive peoples are usually called pacifications, military expeditions, or explorations; with small states, they are called interventions or reprisals; and with internal groups, rebellions or insurrections. Such incidents, if the resistance is sufficiently strong or protracted, may achieve a magnitude that entitles them to the name "war."

In all ages war has been an important topic of analysis. In the latter part of the 20th century, in the aftermath of two world wars and in the shadow of nuclear, biological, and chemical holocaust, more has been written on the subject than ever before. Endeavours to understand the nature of war, to formulate some theory of its causes, conduct, and prevention, are of great importance, for theory shapes human expectations and determines human behaviour. The various schools or theorists are generally aware of the profound influence they can exercise upon life, and their writings usually include a strong normative element, for when accepted by politicians, their ideas can assume the characteristics of self-fulfilling prophecies. The analysis of war may be divided into several categories. Philosophical, political, economic, technological, legal, sociological, and psychological approaches are frequently distinguished. These distinctions indicate the varying focuses of interest and the different analytical categories employed by the theoretician, but most of the actual theories are mixed because war is an extremely complex social phenomenon that cannot be explained by any single factor or through any single approach.

From this analysis it appears that war is an extremely complex phenomenon; that with the progress of science and technology and consequent increases in the destructive power of armaments, war has had an increasingly catastrophic effect on human existence; and, that with the evolution of weapons of mass destruction the task of understanding the nature of war has become more urgent. The analysis recognizes that the major emphasis on past and present attempts to reach a better understanding of war has been directed toward theories and theoreticians. Given special consideration are philosophical, political, economic, technological, legal, sociological and psychological approaches.

At the present time there appears to be an inverse relationship between the extremely rapid increase of the visible catastrophic effects of the powers of wars and the impotence of people to comprehend the depth and breadth of this complex phenomenon. The urgency is immediate — the need for possible benefits to be derived from an unconventional, innovative approach. What is demanded are new, perhaps previously unasked, probably threatening questions. Any or all of these questions ultimately are centered on, begin or end with, those regarding concern of *who*. Other than the recognized theorists, who else may have had thoughts, or words, about war? We should surely consider the ancient mythmakers or mythologists; classical Greek or Roman poets, historians, and dramatists; theologians of any era; modern novelists, essayists, or poets; modern scientists. Who has spoken or written words about war? And what have been the perspectives and interpretations of these men? And who has remained silent, not having spoken or written?

More important are other "who-centered" questions. Who makes or creates war? Who are the participants? Who defines war? Who justifies war, legally, morally, theologically? Who sacrifices or suffers in war? Who kills or is killed? How do we judge the consequences of war, or the intentions of war-makers, or both? Who determines the relative value of those caught in the web of war? And more.

Before assembling a sampling of various voices, each speaking to the issues of war, notice this group of familiar, frequently overlapping words. Conflict may mean fight, battle, war; competitive or opposing action; antagonistic state or action. Antagonism usually means actively expressed opposition or hostility. Strife may mean bitter, sometimes violent conflict or dissension; fight, struggle; exertion, contention, or earnest endeavor for superiority. War may mean a state of usually declared armed hostile conflict between states or nations; a state of hostility, conflict, or antagonism; struggle or competition between opposing forces or for a particular end.

Also, we typify religious wars, political wars, class wars, racial wars, national wars, civil wars, revolutionary wars, tribal wars, sectarian wars, long and short wars, hot and cold wars, war on drugs, war on poverty, war on terror, war of ideas or words, war between the sexes — and perhaps more.

In anticipating what Nietzsche may have had to say about war, it is particularly persuasive that the honor of speaking first be given to Hesiod, the Greek poet, believed to have lived around 700 to 800 BC. It is not difficult to recognize Nietzsche's strong attraction to classical Greek literature — pre-Socratic philosophers, poets, historians, and dramatists, as well as to pre-Hellenic mythology.

Of Hesiod's two extant epic poems, *Theogony* recounts the history of Greek gods and goddesses. It also relates the battles between the Olympians, under the leadership of Zeus, and the Titans, whose leader was Cronus. In this epic, Hesiod further develops the narrative of mythical family relations, of the struggles between male gods and female goddesses.

In Hesiod's other epic, *The Works and Days*, the author is again the storyteller, and an overarching theme remains power struggles involving both mortals and immortals. The following beginning lines from this extraordinary narrative provide some of the richest insights into conflict, war, battle, struggles, etc., in all of Western literature:

> *Muses, who from Pieria give glory through singing,*
>
> *come to me, tell of Zeus, your own father,*
>
> *sing his praises, through whose will*
>
> *mortal men are named in speech or remain unspoken.*
>
> *Men are renowned or remain unsung*
>
> *as great Zeus wills it.*
>
> *For lightly he makes strong,*
>
> *and lightly brings strength to confusion,*
>
> *lightly diminishes the great man,*
>
> *uplifts the obscure one,*
>
> *lightly the crooked man he straightens,*
>
> *withers the proud man,*
>
> *he, Zeus, of the towering thunders,*
>
> *whose house is highest.*
>
> *Hear me, see me, Zeus: hearken:*
>
> *Direct your decrees in righteousness.*
>
> *To you, Perses, I would describe*

the true way of existence.

It was never true that there was only one kind

of strife. There have always

been two on earth. There is one

you could like when you understand her.

The other is hateful. The two Strifes

have separate natures.

There is one Strife who builds up evil war,

and slaughter.

She is harsh; no man loves her, but under compulsion

and by will of the immortals men

promote this rough Strife.

But the other one was born

the elder daughter of black Night.

The son of Kronos, who sits on high and

dwells in the bright air,

set her in the roots of the earth and among men;

she is far kinder.

She pushes the shiftless man to work,

for all his laziness.

A man looks at his neighbor, who is rich:

then he too

wants work; for the rich man presses on with

his plowing and planting

and the ordering of his state.

So the neighbor envies the neighbor

who presses on toward wealth. Such Strife

is a good friend to mortals.

Then potter is potter's enemy, and

craftsman is craftsman's

rival; tramp is jealous of tramp,

and singer of singer.

So you, Perses, put all this firmly away

in your heart,

nor let that Strife who loves mischief

keep you from working

as you listen at the meeting place

to see what you can make of

the quarrels (H, 19, 21).

Later in this book when we approach Nietzsche for some possible thoughts regarding war, this early distinction made by Hesiod may be of inestimable value. Until then, the important translator and interpreter of Nietzsche's writings, R. J. Hollingdale, commenting on a passage from Nietzsche's *Human, All Too Human*, writes:

> The idea behind this passage is of course that, to put it at its broadest, life is a contest, and that the agonistic character of human existence extends into fields where one would not necessarily expect to find it. That the *agon* was the model for all human activity had already suggested itself to Nietzsche during his studies of Greek culture and civilization, where it seemed to him to stand out clearly as a fact. His conclusions had been summarized in "Homer's Contest", the fifth of the *Five Prefaces to Five Unwritten Books* (1872), where he drew attention to the opening of Hesiod's *Works and Days* — the passage in which Hesiod says there are two Eris-goddesses on earth, one the promoter of war, the other "much better one" the promoter of competition — and commented: "The whole of Greek antiquity thinks of spite and envy otherwise than we do, and judges as does Hesiod, who first designates as evil that Eris who leads men against one another to a war of extermination, and then praises another Eris as good who, as jealousy, spite, envy, rouses men to deeds, but not to deeds of war but deeds of *contest*." The conception thus summarized in an unpublished manuscript achieved publication in *Human, All Too Human* in the form: "The Greek artists, the tragedians for example, poetized in order to conquer; their entire art is inconceivable without contest: Hesiod's good Eris, ambition, gave their genius its wings" (*HA*, 170).

This agonistic interpretation of culture is extended, naturally enough, to include the realm of philosophy, and it is done so in a very whole-hearted and comprehensive way: the Greek philosophers, Nietzsche writes, "had a sturdy faith in themselves and their "truth" and with it they overthrew all their contemporaries and predecessors; every one of them was a violent and warlike *tyrant*" (*N*, 78, 79).

Homer, to whom are attributed the two greatest epic poems of ancient Greece, *The Iliad* and *The Odyssey*, was closely contemporaneous with Hesiod. These two lengthy narratives chronicle the protracted Trojan War, a 70-year war between the Greeks and the Trojans, brought on by the abduction of Helen by Paris and ended with the destruction of Troy. The two epic poems deal with gods and goddesses and with Greek heroes. They recount the disastrous consequences of this war, this Greek "victory". In the Introduction to his translation of Hesiod's works, Richmond Lattimore writes, "Heroic legend at the outset leads into myth of cataclysm."

We may be reasonably certain that Nietzsche was familiar with, and influenced by, these two poets. Some of the Greek philosophers, the so-called pre-Socratics, a couple of centuries after Hesiod and Homer, took up the idea of conflict or strife. The particular favorite of Nietzsche's was Heraclitus (c. 540–480 BC). Only a few short fragments of his have survived, usually quoted and attributed to him by later writers. According to Heraclitus, the universe is best understood as what he called *Logos*, meaning something like the controlling principle of the universe, and translated variously as word, order, discourse, pattern, rationale, reason, and more. The Logos, or Word, is revealed as opposites, pairs which are unified by interdependence, but which exist in a state of constant strife.

In Heraclitus' philosophical account, the idea of strife, of struggle, of earnest endeavor for superiority, of sometimes bitter conflict, is extended to explain the relationship, the underlying connection between the opposites which make up the universe — the Logos. Day and night, light and dark, heat and cold — all of the pairs of opposites — exist in this unity, the result of which is a kind of dynamic equilibrium. This constitutes the orderly natural universe. Although philosophical in his attempt to interpret the basic characteristics of nature, Heraclitus' fragments suggest an element of the poetic. And, apparently, he was scorned and was not popular in his own time.

Reflections on war, conflict, strife, competition — from Greek poets, Greek philosophers, and Greek historians. Herodotus (c. 485–425 BC) was the first major Greek historian. He was telling the story of the 27 years of wars between Greece and Persia, giving special attention to the human characters and their actions, their motives, their person-alities, their characters. The reader of Herodotus is made unavoidably aware of certain human qualities which stood out sharply in much of Greek literature of the times — hubris, cruelty, folly, impiety. This poi-gnant passage is from his *Histories*, "In peace children inter their par-ents; war violates the order of nature and causes parents to inter their children."

Thucydides (c. 460–424 BC) is considered the greatest of the an-cient Greek historians. Younger than Herodotus, he was the author of *History of the Peloponnesian War*, and was the contemporary of playwrights Sophocles and Euripides, as well as Socrates. Since we are probing the thinking that has taken place regarding war, and given the status of Thucydides in Greek literature, it seems appropriate to consult again the Encyclopaedia Britannica:

> In a prefatory note near the beginning of the *History*, Thucydides speaks a little of the nature of his task and of his aims. It was difficult, he says, to arrive at the truth of the speeches made — whether he heard them himself or received a report from others — and of the actions of the war. For the latter, even if he himself observed a particular battle, he made as thorough an enquiry as he could — for he realized that eyewitnesses, either from faulty memory or from bias, were not always reliable.
>
> He wrote the speeches out of his own words, appropriate to the occasion, keeping as closely as possible to the general sense of what had actually been said. He could never have omitted them, for it is through the speeches that he explains the mo-tives and ambitions of the leading men and states; and this, the study of the human mind in time of war, is one of his principal aims. (The omission of speeches from the last book is a great loss and is caused, no doubt, by the difficulty he had in getting information about Athens at this period.) He avoided, he says, all "story-telling" (this is a criticism of Herodotus), and his work might be the less attractive in consequence;
>
> but I have written not for immediate applause but for posterity, and I shall be content if the future student of these events, or of other similar events which are likely in human nature to occur in after ages, finds my narrative of them useful.
>
> This is all that he expressly tells of his aim and methods. More-over, in the course of his narrative (except for the pestilence of 430 and his command in 424) he never gives his authority for a statement. He does not say which of the speeches he actu-ally heard, which of the other campaigns he took part in, what

places he visited, or what persons he consulted. Thucydides insisted in doing all the work himself; and he provides, for the parts he completed, only the finished structure, not the plans or the consultations.

Authority of his work. He kept to a strict chronological scheme, and, where it can be accurately tested by the eclipses that he mentions, it fits closely. There are also a fair number of contemporary documents recorded on stone, most of which confirm his account both in general and in detail. There is the silent testimony of the three historians who began where he left off, not attempting, in spite of much independence of opinion, to revise what he had already done, not even the last book, which he clearly did not complete. Another historian, Philistus, a Syracusan who was a boy during the Athenian siege of his city, had little to alter or to add to Thucydides' account in his own *History of Sicily*. Above all, there are the contemporary political comedies of Aristophanes — a man about 15 years younger than Thucydides with as different a temper and writing purpose as could be — which remarkably reinforces the reliability of the historian's dark picture of Athens at war. The modern historian of this war is in much the same position as the ancient; he cannot do much more than translate, abridge, or enlarge upon Thucydides.

For Thucydides kept rigidly to his theme: the history of a war — that is, a story of battles and sieges, of alliances hastily made and soon broken, and, most important, of the behaviour of peoples as the war dragged on and on, of the inevitable "corrosion of the human spirit." He vividly narrates exciting episodes and carefully describes tactics on land and sea. He gives a picture, direct in speeches, indirect in the narrative, of the ambitious imperialism of Athens — controlled ambition in Pericles, reckless in Alcibiades, debased in Cleon — ever confident that nothing was impossible for them, resilient after the worst disaster. He shows also the opposing picture of the slow steadiness of Sparta, sometimes so successful, at other times so accommodating to the enemy.

From a different part of the world, and from a slightly later era — around the early 4th century BC — we have the Chinese author, Suntzu, and his great Chinese classic, *The Art of War*. This work has been recognized as "one of the earliest known compilations on the subject of war and strategy." One brief, often-quoted, passage offers another individual's perspective — "A military operation involves deception. Even though you are competent, appear to be incompetent. Though effective, appear to be ineffective."

Move ahead to the 1st century BC and to Europe, the heir to Greek thought. To Marcus Tullius Cicero (106–43 BC), Roman statesman, orator, and author are attributed these words, "Endless money forms the sinews of war." Here we have another perspective.

Fast forward to the Renaissance, to the early 16th century AD, to the famous Florentine political philosopher, Niccolo Machiavelli (1469-1527), whose primary interest was in formulating a theory concerning the nature and dynamics of power — particularly the power of the ruler, or rulers. In his popular work, *The Prince*, his revolutionary position is this. He assumes that power is not a means to some higher end such as justice, happiness, freedom, etc. Rather, power itself is the end and as such requires no further consideration or debate. The concern of the political philosopher, or politician, is to inquire and elaborate what are the best means of acquiring, retaining, and expanding power. That was Machiavelli's task. And he assigns war as the major means. In *The Prince*, under the heading "Constant Readiness for War", he writes this:

> A prince should therefore have no other aim or thought, nor take up any other thing for his study, but war and its organisation and discipline, for that is the only art that is necessary to one who commands, and it is of such virtue that it not only maintains those who are born princes, but often enables men of private fortune to attain to that rank (*GPT*, 291).

One further consideration of Machiavelli's views regarding power and war should be noted, inasmuch as this connection will also probably be of major interest when we consult Nietzsche on the subject of war. That is, the questions that may arise when giving attention to the possible relation of religion to war. Briefly, the author of the book *Great Political Thinkers, Plato to the Present*, William Ebenstein, comments as follows:

> Machiavelli's views on morals and religion illustrate his belief in the supremacy of power over other social values. He has no sense of religion as a deep personal experience, and the mystical element in religion — its supranatural and suprarational character — is alien to his outlook. Yet he has a positive attitude toward religion; albeit his *religion* becomes a *tool of influence and control* in the hands of the ruler over the ruled. Machiavelli sees in religion the poor man's reason, ethics, and morality put together, and "where religion exists it is easy to introduce armies and discipline." From his reading of history, however, Machiavelli is led to the generalization that "as the observance of divine institutions is the cause of the greatness of republics, so the disregard of them produces their ruin." The fear of the prince, Machiavelli adds, may tempo-

rarily supplant the fear of God, but the life of even an efficient ruler is short.

The role of religion as a mere instrument of political domination, cohesion, and unity becomes even clearer in Machiavelli's advice that the ruler support and spread religious doctrines and beliefs in miracles *that he knows to be false.* The main value of religion to the ruler lies in the fact that it helps him to keep the people "well conducted and united," and from this viewpoint of utility it makes no difference whether he spreads among them true or false religious ideas and beliefs.

Machiavelli's interest in Christianity is not philosophical or theological, but purely pragmatic and political. He is critical of Christianity because "it glorifies more the humble and contemplative men than the men of action," whereas the Roman pagan religion "deified only men who had achieved great glory, such as commanders of republics and chiefs of republics." Christianity idealizes, Machiavelli charges, "humility, lowliness, and a contempt for worldly objects," as contrasted with the pagan qualities of grandeur of soul, strength of body, and other qualities that "render men formidable." The only kind of strength that Christianity teaches is fortitude of soul in suffering; it does not teach the strength that goes into the achievement of great deeds. (*GPT*, 287)

Ebenstein adds this later:

The ultimate commentary on *The Prince* is not in logic or morality, but in experience and history. Machiavelli laid bare the springs of human motivation with consummate and ruthless insight. It is hard to tell, in reading him, whether we shudder more when he tells the truth about human conduct, or more when he distorts the truth. He has unwittingly perhaps, rendered a service to mankind by pointing out the abysmal depth of demoralization and bestiality into which men will sink for the sake of power (*GPT*, 291).

By the time of the emergence of the next major European "philosopher of power", Thomas Hobbes (1588–1679), the dominant twin issues of power and war, together with religion, remained. The early modern age, into which Hobbes was cast, was a period of long and bitter civil and religious conflict and war. This provided the context for his innovative, complex, comprehensive, seminal philosophy. His efforts were directed mainly toward politics and ethics, but extended into religion, psychology, physiology, and language.

Hobbes' theory of politics rests largely on his major assumption regarding the nature of man. Man naturally is combative and competitive. This passage from his major work, *Leviathan*, makes his view clear:

> So that, in the first place, I put for a general inclination of all mankind a perpetual and restless desire of power after power that ceases only in death. And the cause of this is not always that a man hopes for a more intensive delight than he has already attained to, or that he cannot be content with a moderate power, but because he cannot assure the power and means to live well which he has present without the acquisition of more. And from hence it is that kings, whose power is greatest, turn their endeavors to the assuring it at home by laws or abroad by wars; and when that is done, there succeeds a new desire — in some, of fame from new conquests; in others, of ease and sensual pleasure; in others, of admiration or being flattered for excellence in some art or other ability of the mind.

> Competition of riches, honor, command, or other power inclines to contention, enmity, and war, because the way of one competitor to the attaining of his desire is to kill, subdue, supplant, or repel the other (*AE*, 179).

On the basis of this description of mankind, Hobbes further develops his theory of the so-called "state of nature." In this "state of nature", man's original condition was one of war. Arguing for the necessity of a strong government, Hobbes makes his argument:

> Hereby it is manifest that, during the time men live without a common power to keep them all in awe, they are in that condition which is called war, and such a war as is of every man against every man. For war consists not in battle only, or the act of fighting, but in a tract of time wherein the will to contend by battle is sufficiently known; and therefore the notion of *time* is to be considered in the nature of war, as it is in the nature of weather. For as the nature of foul weather lies not in a shower or two of rain but in an inclination thereto of many days together, so the nature of war consists not in actual fighting but in the known disposition thereto during all the time there is no assurance to the contrary.

> In such condition there is no place for industry, because the fruit thereof is uncertain: and consequently no culture of the earth; no navigation nor use of the commodities that may be imported by sea; no commodious building; no instruments of moving and removing such things as require much force; no knowledge of the face of the earth; no account of time; no arts; no letters; no society; and, which is worst of all, continual fear and danger of violent death; and the life of man solitary, poor, nasty, brutish, and short.

To this war of every man against every man, this also is consequent: that nothing can be unjust. The notions of right and wrong,
justice and injustice, have there no place. Where there is no common power, there is no law; where no law, no injustice. Force and
fraud are in war the two cardinal virtues. (*AE*, 181, 182)

Two additional aspects of Hobbes' discussion of power and war are
religion and the recognition that humans require more understanding
of the nature of war. Hobbes apparently had no religion. However, he
approached religion psychologically, reflecting his experiences of living
through great religious upheaval and violence. Typically, with his insistence on the necessity of affirming clear definitions, he defined religion
in general as follows, "*Fear* of power invisible, feigned by the mind, or
imagined from tales publicly allowed, *religion*; not allowed, *superstition*.
And when the power imagined is truly such as we imagine, *true religion*."
Hobbes viewed religions and churches from a standpoint of the serious
dangers of their being a source of civil disobedience or as a pretext for
civil war, which kind of war he considered the worst.

And finally, in answering the questions raised regarding the value,
the utility, the necessity of knowledge — which for Hobbes meant philosophy — he writes, "The end of knowledge is power. . . but the utility
of moral and civil philosophy is to be estimated, not so much by the
commodities we have by knowing these sciences, as by the calamities
we receive from not knowing them." And the greatest calamity was war.
The purpose of knowledge is power. The consequences of not knowing, of ignorance, is fear and weakness, vulnerability to exploitation.
And Hobbes was much concerned with understanding ignorance, not
merely noting that it was the absence of knowledge. It seems worthwhile to note his admonitions. In his *Leviathan*:

Want of science — that is, ignorance of causes — disposes, or
rather constrains, a man to rely on the advice and authority of
others. For all men whom the truth concerns, if they rely not on
their own, must rely on the opinion of some other whom they
think wiser than themselves, and see not why he should deceive
them.

Ignorance of the signification of words, which is want of understanding, disposes men to take on trust, not only the truth they
know not, but also the errors and, which is more, the nonsense
of them they trust; for neither error, nor nonsense can, without a
perfect understanding of words, be detected.

From the same it proceeds that men give different names to one and the same thing from the difference of their own passions: as they that approve a private opinion call it opinion, but they that mislike it, heresy; and yet heresy signifies no more than private opinion, but has only a greater tincture of choler.

Ignorance of natural causes disposes a man to credulity, so as to believe many times impossibilities; for such know nothing to the contrary but that they may be true, being unable to detect the impossibility. And credulity, because men like to be hearkened unto in company, disposes them to lying, so that ignorance itself without malice is able to make a man both to believe lies and tell them — and sometimes also to invent them.

Anxiety for the future time disposes men to inquire into the causes of things, because the knowledge of them makes men the better able to order the present to their best advantage.

And they that make little or no inquiry into the natural causes of things, yet from the fear that proceeds from the ignorance itself of what it is that has the power to do them much good or harm, are inclined to suppose, and feign unto themselves, several kinds of powers invisible and to stand in awe of their own imaginations; and in time of distress to invoke them; as also in the time of an expected good success to give them thanks; making the creatures of their own fancy their gods. By which means it has come to pass that, from the innumerable variety of fancy, men have created in the world innumerable sorts of gods. And this fear of things invisible is the natural seed of that which everyone in himself calls religion, and in them that worship, or fear that power otherwise than they do, superstition.

And this seed of religion having been observed by many, some of those that have observed it have been inclined thereby to nourish, dress, and form it into laws; and to add to it of their own invention any opinion of the causes of future events by which they thought they should be best able to govern others and make unto themselves the greatest use of their powers (*AE*, 180, 181).

A century and a quarter following Hobbes' *Leviathan*, and his concern with the importance, the necessity, of understanding war and its devastating consequences for man, the American, John Adams (1735–1826), wrote these words in 1780 in a letter to his wife, Abigail Adams:

I must study politics and war that my sons may have liberty to study mathematics and philosophy. My sons ought to study mathematics and philosophy, geography, natural history, naval architecture, navigation of commerce, and agriculture, in order to give their children a right to study painting, poetry, music, architecture, statuary, tapestry, and porcelain.

About two years prior to this letter, Abigail Adams had written the following words in a letter to one John Thaxter:

> It is really mortifying, sir, when a woman possessed of a common share of understanding considers the difference of education between the male and female sex, even in those families where education is attended to. . . . Nay why should your sex wish for such a disparity in those whom they one day intend for companions and associates. Pardon me, sir, if I cannot help sometimes suspecting that this neglect arises in some measure from an ungenerous jealousy of rivals near the throne.

At about the same time, she wrote to her husband, "I regret the trifling narrow contracted education of the females of my own country." Educated, perhaps Abigail Adams would also have studied politics and war.

A very different approach to war is taken by the German philosopher, G.W.F. Hegel (1770–1831). Hegel's interpretation and evaluation of war — within the context of his encyclopedic, abstruse, metaphysical, logical, historical, authoritarian, nationalistic philosophy — is not easy to decipher. The best statements of his ideas about war are to be found in his *Philosophy of Law* (1821). Hegel's focus is not on the individual, as with either Machiavelli or Hobbes, but rather on the nation-state. All three develop their theories around the central idea of power. For Hegel, war is an instrument of national policy, the goal of which is victory — power over another nation. He stated, "If states disagree and their particular wills cannot be harmonized, the matter can only be settled by war."

Commenting on Hegel's analysis and justification of war, William Ebenstein writes:

> However, this consequence does not alarm him. On the contrary. Wars are wholesome, he argues, for a number of reasons. First, "corruption in nations would be the product of prolonged, let alone "perpetual peace." The trouble with peace, Hegel thinks, is that men stagnate in it. Their idiosyncrasies become more fixed and ossified, and war is desirable to keep the body politic healthy through proper stimulation. In addition, Hegel stresses the salutary impact of war on national unity: "As a result of war, nations are strengthened, but peoples involved in civil strife also acquire peace at home through making war abroad" (*GPT*, 607).

In Hegel's view of world history, wars are the means of deciding which of the conflicting nations possess the greatest power, and therefore the greatest value. Victory in war ultimately decides which nation's values are most excellent. War is real and it is rational.

Of the same generation as Hegel, but more well known for his views on war, is the Prussian general and military strategist, Carl von Clausewitz (1780–1831). Clausewitz's fame rests on his book, *On War*, and his basic ideas appear to have remained generally accepted. Prior to his own military experiences, he was educated in literature and philosophy, and his analysis of war reflects these interests. He "tried to analyze the workings of military genius by isolating the factors that decide success in war." He was interested less in technical matters and more in psychological considerations. He is often quoted as saying, "War is not merely a political act, but also a political instrument, a continuation of political relations, a carrying out of the same by other means." In Clausewitz's view, war is never an end in itself, but always only a means. That is, always a means to attain a position of power — as is all politics.

And finally, closer to our own time are two of the most famous and influential figures of the 20th century. In *Great Political Thinkers*, William Ebenstein wrote:

> In 1932 the League of Nations and the International Institute of Intellectual Cooperation asked Albert Einstein, the greatest physicist since Newton, to choose a significant issue of the time, and to discuss it with a person to be selected by him. Einstein chose the problem: "Is there a way of delivering mankind from the menace of war?" as the most important issue of humanity, and he selected Freud as the person most qualified to illuminate its cause and cure. This exchange of letters was published by the League of Nations under the title of *Why War?* in 1933. By that time the establishment of the Nazi regime in Germany had made the question one of tragic timeliness (*GPT*, 837).

Einstein's letter deserves inclusion here in its entirety, especially inasmuch as the questions he puts to Freud point the direction to setting the stage for Nietzsche to be consulted. Understanding war must continue to depend on insights proposed by psychologists. Nietzsche would agree. Here is Einstein's letter:

Caputh near Potsdam,
30th July, 1932.

Dear Professor Freud,

The proposal of the League of Nations and its International Institute of Intellectual Co-operation at Paris that I should invite a person to be chosen by myself, to a frank exchange of views on any problem that I might select affords me a very welcome opportunity of conferring with you upon a question which, as things now are, seems the most insistent of all the problems civilisation has to face. This is the problem: Is there any way of delivering mankind from the menace of war? It is common knowledge that, with the advance of modern science, this issue has come to mean a matter of life and death for civilisation as we know it; nevertheless, for all the zeal displayed, every attempt at its solution has ended in a lamentable breakdown.

I believe, moreover, that those whose duty it is to tackle the problem professionally and practically are growing only too aware of their impotence to deal with it, and have now a very lively desire to learn the views of men who, absorbed in the pursuit of science, can see world-problems in the perspective distance lends. As for me, the normal objective of my thought affords no insight into the dark places of human will and feeling. Thus, in the enquiry now proposed, I can do little more than seek to clarify the question at issue and, clearing the ground of the more obvious solutions, enable you to bring the light of your far-reaching knowledge of man's instinctive life to bear upon the problem. There are certain psychological obstacles whose existence a layman in the mental sciences may dimly surmise, but whose interrelations and vagaries he is incompetent to fathom; you, I am convinced, will be able to suggest educative methods, lying more or less outside the scope of politics, which will eliminate these obstacles.

As one immune from nationalist bias, I personally see a simple way of dealing with the superficial (i.e. administrative) aspect of the problem; the setting up, by international consent, of a legislative and judicial body to settle every conflict arising between nations. Each nation would undertake to abide by the orders issued by this legislative body, to invoke its decision in every dispute, to accept its judgments unreservedly and to carry out every measure the tribunal deems necessary for the execution of its decrees. But here, at the outset, I come up against a difficulty; a tribunal is a human institution which, in proportion as the power at its disposal is inadequate to enforce its verdicts, is all the more prone to suffer these to be deflected in extrajudicial pressure. This is a fact with which we have to reckon; law and might inevitably go hand in hand, and juridical decisions approach more nearly the ideal justice demanded by the community (in whose name and interests these verdicts are pronounced) in so far as the community has effective power to compel respect of its juridical ideal. But at present we are far from possessing any supranational organisation competent to render verdicts of incontestable authority and enforce absolute submission to the execution of its verdicts. Thus I am led to my first axiom: the quest of international security involves the unconditional surrender by every nation, in a certain measure, of its liberty of action, its sovereignty that is to say, and it is clear beyond all doubt that no other road can lead to such security.

The ill-success, despite their obvious sincerity, of all the efforts made during the last decade to reach this goal leaves us no room to doubt that strong psychological factors are at work, which paralyse these efforts. Some of these factors are not far to seek. The craving for power which characterises the governing class in every nation is hostile to any limitation of the national sovereignty. This political power-hunger is wont to batten on the activities of another group, whose aspirations are on purely mercenary, economic lines. I have specially in mind that small but determined group, active in every nation, composed of individuals who, indifferent to social considerations and restraints, regard warfare, the manufacture and sale of arms, simply as an occasion to advance their personal interests and enlarge their personal authority.

But recognition of this obvious fact is merely the first step towards an appreciation of the actual state of affairs. Another question follows hard upon it: How is it possible for this small clique to bend the will of the majority, who stand to lose and suffer by a state of war, to the service of their ambitions? (In speaking of the majority, I do not exclude soldiers of every rank who have chosen war as their profession, in the belief that they are serving to defend the highest interests of their race, and that attack is often the best method of defence.) An obvious answer to this question would seem to be that the minority, the ruling class at present, has the schools and press, usually the Church as well, under its thumb. This enables it to organise and sway the emotions of the masses, and make its tool of them.

Yet even this answer does not provide a complete solution. Another question arises from it: How is it these devices succeed so well in rousing men to such wild enthusiasm, even to sacrifice their lives? Only one answer is possible. Because man has within him a lust for hatred and destruction. In normal times this passion exists in a latent state, it emerges only in unusual circumstances; but it is a comparatively easy task to call it into play and raise it to the power of a collective psychosis. Here lies, perhaps, the crux of all the complex factors we are considering, an enigma that only the expert in the lore of human instincts can resolve.

And so we come to our last question. Is it possible to control man's mental evolution so as to make him proof against the psychoses of hate and destructiveness? Here I am thinking by no means only of the so-called uncultured masses. Experience proves that it is rather the so-called "Intelligentsia" that is most apt to yield to these disastrous collective suggestions, since the intellectual has no direct contact with life in the raw, but encounters it in its easiest, synthetic form — upon the printed page.

To conclude: I have so far been speaking only of wars between nations; what are known as international conflicts. But I am well aware that the aggressive instinct operates under other forms and in other circumstances. (I am thinking of civil wars, for instance, due in earlier days to religious zeal, but nowadays to social factors; or, again, the persecution of racial minorities.) But my insistence on what is the most typical, most cruel and extravagant form of conflict between man and man was deliberate, for here we have the best occasions of discovering ways and means to render all armed conflicts impossible.

I know that in your writings we may find answers, explicit or implied, to all the issues of this urgent and absorbing problem. But it would be of the greatest service to us all were you to present the problem of world peace in the light of your most recent discoveries, for such a presentation well might blaze the trail for new and fruitful modes of action.

Yours very sincerely,

Einstein (*GPT*, 851, 852, 853)

Rather than reproduce Freud's lengthy reply, Ebenstein's summary is preferable:

> Applying his two basic concepts of erotic and destructive impulses, Freud came to the conclusion that "there is no likelihood of our being able to suppress humanity's aggressive tendencies." The problem of eliminating war, Freud argues, cannot be tackled directly by seeking to abolish the aggressive impulses in man, but must be tackled indirectly by diverting these impulses into channels other than warfare and by strengthening the emotional ties that bring men together in common efforts and enjoyments. Freud also holds that the organization of society along rational principles indirectly strengthen the forces of peace. All such methods, however, he is bound to admit, "conjure up an ugly picture of mills that grind so slowly that, before the flour is ready, men are dead of hunger."

> Yet, owing to two facts, Freud is not wholly pessimistic about the possibility of eliminating war: first, man's progressive ability to master intellectually his instinctual life, and second, "a well-

founded dread of the form that future wars will take." Civilization is essentially the process of taming and domesticating man's
instincts, and Freud does not know how the next phase of that
process will come about. But he ends with the not altogether
hopeless note that "meanwhile we may rest on the assurance that
whatever makes for cultural development is working also against
war" (*GPT*, 837, 838).

Of the issues and questions raised by Einstein, perhaps the most
provocative is this. Referring to the governing class of every nation:

> How is it possible for this small clique to bend the will of the
> majority, who stand to lose and suffer by a state of war, to the
> service of their ambitions? (In speaking of the majority, I do not
> exclude soldiers of every rank who have chosen war as their pro-
> fession, in the belief that they are serving to defend the highest
> interests of their race, and that attack is often the best method of
> defence.) An obvious answer to this question would seem to be
> that the minority, the ruling class at present, has the schools and
> press, usually the Church as well, under its thumb. This enables
> it to organise and sway the emotions of the masses, and make its
> tool of them (*GPT*, 852).

Perhaps we may be able to put the question to Nietzsche with the
hope, or anticipation, of some success. *Why War?*

Part Two

It would be difficult not to imagine that Friedrich Nietzsche, had his creative life endured past his forty-fourth year, would have directed his critical capacities, his genius, toward *war*. The first part of this book was intended to assemble a brief background upon which to sketch a portrait of the person who himself laid a broad base of ideas upon which we may dare to project images of a possible new and bold interpretation of war. Nietzsche, we may believe, would applaud such an effort.

In Part One of this book, it seemed worthwhile to remind ourselves that there are a group of familiar, frequently overlapping words, and attention was called to these:

> Conflict may mean fight, battle, war; competitive or opposing action; antagonistic state or action. Antagonism usually means actively expressed opposition or hostility. Strife may mean bitter, sometimes violent conflict or dissension; fight, struggle; exertion, contention, or earnest endeavor for superiority. War may mean a state of usually declared armed hostile conflict between states or nations; a state of hostility, conflict, or antagonism; struggle or competition between opposing forces or for a particular end.

Further, it was suggested that we recognize numerous types, or categories, of what we often consider as "wars":

Also, we typify religious wars, class wars, racial wars, national wars, civil wars, revolutionary wars, tribal wars, sectarian wars, long and short wars, hot and cold wars, war on drugs, war on poverty, war on terror, war of ideas or words, war between the sexes — and perhaps more.

And we may identify special interests, or approaches, such as the art of war, the science of war, the psychology of war, the morality of war, the legality of war, the politics of war, the economics of war, and others.

From the limited number of selected voices gathered in Part One to describe, characterize, or interpret war, these words stand out — cataclysmic, deception, catastrophe, ignorance, evil, violent, extermination, hubris, cruelty, folly, force and fraud, ravages, brutality, vulgar, corrosion of the human spirit, violation of nature, an invention, a political act, a political instrument or tool of control. From different eras, in different situations, with different interests, the various writers cited, when called upon, have each contributed to a compelling chorus of catastrophe.

Listening to these voices, individually or collectively, there appears to be about perfect congruence. War is detestable, a curse on humanity. However, the question we would put to Nietzsche remains with no satisfactory answer — *Why War?* "Why" meaning "for what cause, reason, or purpose", or "signalling that we are faced with a baffling problem, an enigma". From among the several men to whom we appealed for some insight on war in Part One of this book, I have chosen these as most likely to have influenced Nietzsche's thinking on the subject, directly or indirectly — Hesiod, Heraclitus, Machiavelli, and Hobbes. A fifth person is J. J. Bachofen. From among these five, the four chosen as having contributed most to Nietzsche's thinking on war — what he said and what he might say — are Hesiod, Heraclitus, Hobbes, and Bachofen. Yes, Nietzsche is best described as a "thinker", and for over a century, countless individuals have continued the process of thinking and rethinking his life, what he said or wrote, what he meant or may have meant.

Briefly, if we consider Nietzsche as a thinker, and a complex and comprehensive thinker, it calls attention to the following. He studied early to become a theologian. He abandoned theology and became a philologist, a philosopher, a poet, a psychologist, a historian, a cultural critic. His major interests or concerns included nature, religion, history, politics, arts, sciences, and myths. The ideas which became dominant and evolved throughout his life were power, values, morality, contraries, and sexuality.

Recognizing the depth and breadth of Nietzsche's abilities, his many perspectives, leads a reader to listen to him when he reflects on being a "deep thinker". Reminded by Martin Heidegger, a devoted disciple to Nietzsche, that in attempting to understand the direction of Nietzsche's thinking, one must never ignore, or minimize, the idea of *reversal*. Heidegger was familiar with Nietzsche's report that frequently, rather than it being the case that "I think thoughts," the experience is, "thoughts think me." Heidegger himself wrote what was possibly his most exciting and provocative book, *What Is Called Thinking?* And this was more or less in response to Nietzsche's many queries on the subject. The most memorable quotation from Heidegger's book is, "Most thought-provoking in our thought-provoking time is that we are still not thinking."

In a passage in Nietzsche's last book, *Ecce Homo*, in which he is reflecting on the concept "inspiration", he vividly describes his own experience of what the process of thinking, or part of the process, must frequently have been for him:

> Has anyone at the end of the nineteenth century a distinct conception of what poets of strong ages called *inspiration*? If not, I will describe it. — If one had the slightest residue of superstition left in one, one would hardly be able to set aside the idea that one is merely incarnation, merely mouthpiece, merely medium of overwhelming forces. The concept of revelation, in the sense that something suddenly, with unspeakable certainty and subtlety, becomes *visible*, audible, something that shakes and overturns one of the depths, simply describes the fact. One hears, one does not seek; one takes, one does not ask who gives; a thought flashes up like lightning, of necessity, unfalteringly formed — I have never had any choice. An ecstasy whose tremendous tension sometimes discharges itself in a flood of tears, while one's steps now involuntarily rush along, now involuntarily lag; a complete being outside of oneself with the distinct consciousness of a multitude of subtle shudders and trickles down to one's toes; a depth of happiness in which the most painful and gloomy things appear, not as an antithesis, but as conditioned, demanded, as a *necessary* colour within such a superfluity of light; an instinct for

rhythmical relationships which spans forms of wide extent — length, the need for a *wide-spanned* rhythm is almost the measure of the force of inspiration, a kind of compensation for its pressure and tension . . .

Everything is in the highest degree involuntary but takes place as in a tempest of a feeling of freedom, of absoluteness, of power, of divinity . . . The involuntary nature of image, of metaphor is the most remarkable thing of all; one no longer has any idea of what is image, what metaphor, everything presents itself as the readiest, the truest, the simplest means of expression. It really does seem, to allude to a saying of Zarathustra's, as if the things themselves approached and offered themselves as metaphors (— "here all things come caressingly to your discourse and flatter you; for they want to ride upon your back. Upon every image you here ride to every truth. Here, the words and word-chests of all existence spring open to you; all existence here wants to become words, all becoming here wants to learn speech from you — "). This is *my* experience of inspiration; I do not doubt that one has to go back thousands of years to find anyone who could say to me "it is mine also" (*EH*, 102, 103).

Recognize that if Nietzsche, as we might wish, had addressed the meaning and significance of war with similar intensity and perseverance as he displayed in his critical interpretation and evaluation of Christianity, he would necessarily, I believe, have brought his major intellectual perspectives, interests, and ideas to the subject in full force. I will suggest only, as we attempt to tease from him some possible incipient views, that he would have seen war — in the sense of "hostile, violent, armed aggression" — much as he saw Christianity — as leading inexorably to the decadence, the deterioration, the degradation, the death and decomposition of Western culture. The two most dangerous inventions, experiments, lies — threatening perhaps the survival of more than we can imagine.

As was suggested in Part One, it is particularly persuasive to begin this task and adventure by revisiting the Greek poet, Hesiod, one of the earliest writers who stands out in Western literature as having impressed and influenced Nietzsche's thinking. Recall that in his epic poem, *Theogony*, Hesiod narrates the mythical struggles between male gods and female goddesses. In the other extant poem, *The Works and Days*, he tells of the power struggles involving both mortals and immortals, further developing the engagement in the struggles between females and males. To reinforce our sense of these remarkable, eloquent words and of the foundational insights into the nature of human conflicts, hear again Hesiod:

> *Muses, who from Pieria give glory through singing,*
>
> *come to me, tell of Zeus, your own father,*
>
> *sing his praises, through whose will*
>
> *mortal men are named in speech or remain unspoken.*
>
> *Men are renowned or remain unsung*
>
> *as great Zeus wills it.*
>
> *For lightly he makes strong,*
>
> *and lightly brings strength to confusion,*
>
> *lightly diminishes the great man,*
>
> *uplifts the obscure one,*
>
> *lightly the crooked man he straightens,*
>
> *withers the proud man,*
>
> *he, Zeus, of the towering thunders,*
>
> *whose house is highest.*
>
> *Hear me, see me, Zeus hearken:*
>
> *Direct your decrees in righteousness,*
>
> *To you, Perses, I would describe*
>
> *the true way of existence.*
>
> *It was never true that there was only one kind*
>
> *of strife. There have always*
>
> *been two on earth. There is one*
>
> *you could like when you understand her.*
>
> *The other is hateful. The two Strifes*
>
> *have separate natures.*
>
> *There is one Strife who builds up evil war,*
>
> *and slaughter.*
>
> *She is harsh; no man loves her, but under compulsion*
>
> *and by will of the immortals men*
>
> *promote this rough Strife.*

But the other one was born

the elder daughter of black Night.

The son of Kronos, who sits on high and

dwells in the bright air,

set her in the roots of the earth and among men;

she is far kinder.

She pushes the shiftless man to work,

for all his laziness.

A man looks at his neighbor, who is rich:

then he too

wants work; for the rich man presses on with

his plowing and planting

and the ordering of his state.

So the neighbor envies the neighbor

who presses on toward wealth. Such Strife

is a good friend to mortals.

Then potter is potter's enemy, and

craftsman is craftsman's

rival; tramp is jealous of tramp,

and singer of singer.

So you, Perses, put all this firmly away

in your heart,

nor let that Strife who loves mischief

keep you from working

as you listen at the meeting place

to see what you can make of

the quarrels (H, 19, 21).

R. J. Hollingdale observes:

The idea behind this passage is of course that, to put it at its broadest, life is a contest, and that the agonistic character of human existence extends into fields where one would not necessarily expect to find it. That the *agon* was the model for all human activity had already suggested itself to Nietzsche during his studies of Greek culture and civilization, where it seemed to him to stand out clearly as a fact. His conclusions had been summarized in "Homer's Contest", the fifth of the *Five Prefaces to Five Unwritten Books* (1872), where he drew attention to the opening of Hesiod's *Works and Days* — the passage in which Hesiod says there are *two* Eris-goddesses on earth, one the promoter of war, the other "much better one" the promoter of competition — and commented: "The whole of Greek antiquity thinks of spite and envy otherwise than we do, and judges as does Hesiod, who first designates as evil that Eris who leads men against one another to a war of extermination, and then praises another Eris as good who, as jealousy, spite, envy, rouses men to deeds, but not to deeds of war but deeds of *contest*." The conception thus summarized in an unpublished manuscript achieved publication in *Human, All Too Human* in the form: "The Greek artists, the tragedians for example, poetized in order to conquer; their entire art is inconceivable without contest. Hesiod's good Eris, ambition, gave their genius its wings" (*HA*, 170).

This agonistic interpretation of culture is extended, naturally enough, to include the realm of philosophy, and it is done so in a very whole-hearted and comprehensive way: the Greek philosophers, Nietzsche writes, "had a sturdy faith in themselves and their 'truth' and with it they overthrew all their contemporaries and predecessors; every one of them was a violent and warlike *tyrant*" (*N*, 78, 79).

The agonistic character of human existence — agonistic being understood as "of, or relating to, or being aggressive, or defensive social interaction (as fighting, fleeing, or submitting) between individuals, usually of the same species" — became an established principle in interpreting human behavior.

No other interpretation of conflict in ancient Greek literature is as foundational as the two Eris-goddesses of Hesiod, one the promoter of war, the other the promoter of competition and contest. Conflict and competition — providing the distinction and basis for interpreting both nature and culture. Respectful of Hobbes' insistence that a first requirement of any argument, analysis, or interpretation is to present clear definitions, I suggest these. Nature may be understood as a creative and controlling force in the universe, and regarded as the source of all animate and inanimate things, as the source of life. Culture may be defined as the superstructure, a continuation or reflection of nature, of the natural substructure, either one of affirmation or of negation of nature.

Early in Nietzsche's development he may have been considering Hesiod's two Eris-goddesses, one the promoter of competition, the other the promoter of war. Nietzsche's notion of benefitting was related to competition. His notion of hurting was related to war. Two opposing possibilities — creation and destruction.

In addition to the sense of agon, as contest or conflict, being the model for all human activity, and of agonistic as being the principle of interpretation for human culture, these cognates were of major significance in Nietzsche's thinking. Protagonist — understood as the principal character in a literary work (as a drama or story) or real event, or as a leader, proponent, or supporter of a cause. Antagonist — understood as one who opposes or contends with another; adversary, opponent. Antagonism — considered as the opposition of a conflicting force. And especially there is agonize — to struggle, suffer agony, anguish over every decision. There is certainly abundant evidence in Nietzsche's writing to indicate that much of his thinking was of this nature. His life was an exemplar of the agon — both protagonist and antagonist. And it is not difficult to see in Hesiod's poetry the seeds of these major ideas of Nietzsche — power, values, morality, contraries, and sexuality.

No one should doubt the importance of Hesiod's poetic interpretation of strife — for future thinkers, and obviously for Nietzsche. And the subject appears even more significantly approximately two centuries after Hesiod in the philosophy of the so-called pre-Socratic philosopher, Heraclitus (c. 540–c. 480 BC). Given what appears to be the case, that from among an assorted group of early philosophers predating Socrates and Plato, Heraclitus and his philosophy of nature captured Nietzsche's imagination the most and influenced his thinking so extraordinarily, he deserves close attention. One could perhaps easily make the case that of all prior philosophers, Heraclitus stands near the top of the list of those influencing not only Nietzsche's thinking but all subsequent philosophical thinking. And this in spite of there being available such a small amount of what is believed to be his own words, usually referred to as "fragments".

According to Heraclitus, the universe is best understood as what he called "Logos", meaning something like the controlling principle or the formative pattern, the rationale. He wrote, "Of this Rationale, which is eternal, men turn out to be ignorant, both before they hear it and when they hear it for the first time. For though all things occur in accordance with the Rationale, they are like novices when they are tested by such words and works as I work out, distinguishing each thing according to its nature and explaining what it is. But such things as they do when they are awake escape other men, just as they forget about what they do when asleep." And this, "Listening not to me but to the Rationale, it is wise to agree that all things are one." This Rationale reveals the unity of all things. Also, this Logos, or Word, is revealed as consisting entirely throughout as pairs of opposites, every pair being itself unified by interdependence, but each existing in a state of constant strife. No member of any pair of opposites is capable of existing on its own, i.e., independently. All of these innumerable pairs that make up the world do exist, but never in a fixed, stable state. Within each pair there is a constant process of change and exchange. They are constantly opposing one another. In the view of Heraclitus, the strife, or conflict, of opposites is the basic fact of existence. Everything in the universe is involved in this eternal process, always involving pairs of opposites. Heraclitus established a view of the world which included constant change and at the same time eternal order, or pattern, within which changes occur. The world does not consist of unrelated things, as some of his contemporaries believed. Neither is it a chaos of haphazard, disconnected events. Heraclitus wrote, "War is father of all," And, "all things happen in accordance with strife." To believe, or desire, it otherwise is folly, to wish for the end of existence. Heraclitus, critical of both Hesiod and Homer, wrote, "Homer deserved to be thrown out from amongst the contestants and beaten" — for he advocated pacifism.

About a sixth of the extant fragments of Heraclitus deal with opposites, and some are especially worth noting. "Over those who step into the same river, different and again different waters flow." "Beginning and end are common on the circumference of a circle." "They would not know the name of Justice if the things [injustices] did not exist." "Sickness makes health pleasant and good; hunger, satiety; weariness, rest." Another kind of polar opposition consists in the regular succession by one thing of its opposite, so that if one perished, so would the other. "The cold things get hot, hot gets cold, wet gets dry, parched gets damp." Or, the alternation between night and day, sun and moon, and numerous others.

Heraclitus also points to another type of opposition, or conflict. Many have to do with the observer, the special interests, perspectives, tastes, etc. The same road is "the road up" to those living below and "the road down" to those living above. Or, the same object may be beautiful to one person, ugly to another. Or, the same thing or action good or right to one person, and bad or wrong to another. Also, these values may be different at different times for the same person regarding the same object or event. The case may be made that every one of these assertions, or examples, of Heraclitus had a major impact on Nietzsche's philosophy.

An important, perhaps frequently overlooked aspect of this eternal order or pattern of existence, the conflict of opposites, needs noting. The notion of "measures" is intrinsic to the Rationale, the formative pattern. Heraclitus wrote, "The sun will not overstep; if he did, Furies, guardian of Justice, would find him out."

Although philosophical speculation about the nature of things, as distinguished from mythology, had existed only less than a century when Heraclitus wrote, these statements of his are steeped in mythological suggestions or meanings. Looking ahead to further consideration of mythological and symbolic sources important in Nietzsche's thinking, it is of interest to elaborate a little on that last quotation of Heraclitus. "The sun will not overstep measure; if he did, Furies, guardian of Justice, would find him out."

Briefly, the sun, the solar symbolism extends in its broadest terms around the male. The male creator, the masculine force, the heroic principle, the sun-king, and much more is associated symbolically and mythologically with the male. The association of the moon with the female, the feminine force, the female creator, etc., together with the sun is a major instance of the symbolism of dualism central to the philosophy of Heraclitus. Any reader of Nietzsche will realize that the same is true of his philosophy. There is good reason to believe that Heraclitus was familiar with this fundamental pair of opposites — male and female. In mythological accounts, the Furies, three females, personified the anger and revenge of the Triple Goddess Demeter — the punishing or scolding Mother. The Furies in classical tradition were understood as a symbol of the impersonal function of justice. And in classical paganism, the spirit of justice was female, reflecting "natural justice". Early Greek philosophy, especially those fragments of Heraclitus, was still intimately entwined with the rich traditions of myths and symbols, as Nietzsche became well aware.

Hesiod, Heraclitus, and centuries later the English philosopher, Thomas Hobbes (1588–1679) — three giant figures in Nietzsche's background, and especially critical in our search for what could have been his help in understanding the questions surrounding the phenomenon of *war*. Hesiod, the poet and storyteller, from whom Nietzsche gathered so much from so few words. Heraclitus, one of the earliest philosophers, the same situation — the power of few words. And Hobbes, a philosopher of many words, appears to have opened up a panorama of exciting, challenging, heretical new ideas with unsuspected possibilities — just the type of thinking that would have later had Nietzsche saying, "I don't think thoughts, thoughts think me." And perhaps he would have said also that he was always constantly and unsuccessfully resisting those thoughts.

Hobbes' reputation and influence as a *thinker* rests primarily on his philosophy of *man*. Around this focus, he developed his identity and an amazing expanse of ideas. His time, the events, and his own life experiences were the crucible for the emergence of this formidable and revolutionary philosopher. As noted earlier, Hobbes was born and lived his long life during tumultuous times of extended and bitter religious and civil conflict and war. No previous philosopher would have been living, thinking, and writing with more constant intimate involvement with human competition and conflict than Hobbes — in both wars of physical violence and destruction and those of conflict and competition of new ideas and words.

Living in the late years of the Renaissance, Hobbes was heir to a legacy of a lengthy period of vigorous artistic and intellectual activity. At the same time, he was a leading figure in the early stages of the modern scientific era — with obvious dire implications for the religious structures of Europe. Early in his educational pursuits, Hobbes apparently discovered the scientific theories of Galileo and Johannes Kepler relating to planetary motion. Soon after, he decided to become a student of classical studies, became a classical scholar, translated Thucydides and later Homer's *Iliad* and *Odyssey*. Almost certainly, Hobbes became familiar with Hesiod and with Heraclitus and their early thoughts regarding strife and the continuous conflict of opposites, as well as other early philosophical theories focusing on nature, on matter, and on motion. His interests became science and philosophy. His defining concentration was on motion — the diversity of motion, the principles of motion. His sweeping claim was that the basic reality of nature consists of "matter in motion" or "bodies in motion". And, his concerns above all others were to understand human bodily motions — man in motion — especially bodily motions involved in cognition.

These two primary interests — philosophy and science — informing each other, together propelled Hobbes both backward and forward in time. The scope of his opportunities for new intellectual possibilities must have seemed unlimited. That he fully embraced these opportunities is not in question. In attempting to put into play a range of ideas that might serve to advance the evolution of our consciousness and understanding regarding the resistant puzzles of war — especially with the possible guidance of Nietzsche — the thinking of Hobbes merits close scrutiny.

To begin, the notion emphasized by Heraclitus that the natural universe is basically ordered in terms of pairs of opposites, unified by interdependence, was adopted uncritically by Hobbes. Examples: presence and absence, natural and acquired, good and evil, simple and complex, appetite and aversion, love and hate, cause and effect, power and weakness, and of course, war and peace — and more. Neither did Hobbes need any persuasion of the ubiquity of strife, conflict, competition, war. He was familiar with the writings of Hesiod, Homer, Thucydides, and he was living, and engaged in constantly recurring violence. Response to his creative and critical writings often meant that his own life was at times in jeopardy, that the charge of heresy was frequently brought against him.

The claim may be made, and adequately defended, that no single philosopher had created a network of ideas more rich in possibilities for further development than Thomas Hobbes — especially for Nietzsche.

My intention is to look at what are the major elements of that network — at least some of them — in anticipating that they may be recognized as a rich source for validating and interpreting Nietzsche's thinking.

Nietzsche believed, Hobbes believed, and earlier Machiavelli believed, if in a more restricted context, that the basic starting point, the center of all philosophy and science was *power*. All other ideas seemed to be generated from, or to refer back to, this base. Machiavelli certainly understood that central to war was power, and the reverse was also the case. So before continuing with Hobbes, we may benefit by a closer look at Machiavelli.

A contemporary during the Renaissance of Leonardo da Vinci, whose creative brilliance was expressed in the arts, in science and engineering, in philosophy and literature, Machiavelli focused his intellectual vitality almost entirely on the nature and dynamics of power, particularly in the study of politics. He indicated little concern with economics or with a broad range of cultural pursuits. Rather, it was the military state and the power, or powers, of the ruler, the prince, that were his focus. Machiavelli's prince appears to have been a single man, whose forces were directed toward coercing submission or inaction. The threat of force, or the use of force, were the bases of the ruler's power.

Reversing the conventional assumption that the power of the ruler is the means to some further defined end, e.g., freedom, justice, protection, etc., Machiavelli, in his popular work *The Prince*, declared that power is the end, in relation to which anything and everything else is always a means. His concerns, as a political philosopher, were to inquire and elaborate the appropriate and effective means of acquiring, maintaining, and increasing the power of the ruler, and that meant of the state. For one who rules, the end remains the same — power. As for the means, they are different, constantly changing and uncertain, often interrupted by chance or fortune. Toward the goal, the prudent ruler will shift with events or circumstances, often exhibiting contradictory choices or means.

Machiavelli does elaborate somewhat on the nature of man and on certain important psychological perspectives. Men, he says, are weak, fickle, changeable in their loyalties. As for concern with these, his subjects, the ruler considers such concerns only in terms of their value for continuing his rule and power. Three particular instruments, or means, stand out in Machiavelli's account — fear, cruelty, and deceit.

A few comments regarding fear — the person in command is fearful of desertion, of possible rebellion, and of conspiracy by those who submit to his power. He is fearful of his enemies, and of the possible loss of power. At the same time, fear is the primary tool for achieving obedience and submission. As for cruelty, in itself it is neither good nor bad. It may be required, or chosen, may seem necessary as a means to maintain or enhance power, and therefore good. Regarding deceit, the prudent ruler will keep his word, unless deceit is necessary. Machiavelli writes, "Men are so simple, so ready to follow the needs of the moment that the deceiver will always find someone to deceive." Deception, either of the ruler's subjects or of his enemies, may be necessary and effective, and therefore good. Good and evil are never absolute and stable, always relative and changing.

In acquiring, maintaining, and expanding his power, the major means of the ruler in the state, according to Machiavelli, is *war*. In *The Prince*, under the heading "Constant Readiness for War", again these words:

> A prince should therefore have no other aim or thought, nor take up any other thing for his study, but war and its organization and discipline, for that is the only art that is necessary to one who commands, and it is of such virtue that it not only maintains those who are born princes, but often enables men of private fortune to attain to that rank (*GPT*, 291).

War, in Machiavelli's view, is more an art than a science.

As to the views of Machiavelli on the subject of the possible relation of religion to war, he is consistent. I would call attention again to this interpretation by William Ebenstein:

> Machiavelli's interest in Christianity is not philosophical or theological, but purely pragmatic and political. He is critical of Christianity because "it glorifies more the humble and contemplative men than the men of action," whereas the Roman pagan religion "deified only men who had achieved great glory, such as commanders of republics and chiefs of republics." Christianity idealizes, Machiavelli charges, "humility, lowliness, and a contempt for worldly objects," as contrasted with the pagan qualities of grandeur of soul, strength of body, and other qualities that "render men formidable." The only kind of strength that Chris-

tianity teaches is fortitude of soul in suffering; it does not teach the strength that goes into the achievement of great deeds (*GPT*, 287).

Machiavelli's views on morals and religion illustrate his belief in the supremacy of power over other social values. He has no sense of religion as a deep personal experience, and the mystical element in religion — its supranatural and suprarational character — is alien to his outlook. Yet he has a positive attitude toward religion; albeit his *religion* becomes a *tool of influence and control* in the hands of the ruler over the ruled. Machiavelli sees in religion the poor man's reason, ethics, and morality put together, and "where religion exists it is easy to introduce armies and discipline." From his reading of history, however, Machiavelli is led to the generalization that "as the observance of divine institutions is the cause of the greatness of republics, so the disregard of them produces their ruin." The fear of the prince, Machiavelli adds, may temporarily supplant the fear of God, but the life of even an efficient ruler is short.

The role of religion as a mere instrument of political domination, cohesion, and unity becomes even clearer in Machiavelli's advice that the ruler support and spread religious doctrines and beliefs in miracles *that he knows to be false*. The main value of religion to the ruler lies in the fact that it helps him to keep the people "well conducted and united," and from this viewpoint of utility it makes no difference whether he spreads among them true or false religious ideas and beliefs (*GPT*, 287).

And Ebenstein's summary, useful to remember later when we engage Nietzsche on power and war, reads:

> The ultimate commentary on *The Prince* is not in logic or morality, but in experience and history. Machiavelli laid bare the springs of human motivation with consummate and ruthless insight. It is hard to tell, in reading him, whether we shudder more when he tells the truth about human conduct, or more when he distorts the truth. He has unwittingly perhaps, rendered a service to mankind by pointing out the abysmal depth of demoralization and bestiality into which men will sink for the sake of power (*GPT*, 291).

Hobbes built his extensive philosophy, especially his theory of politics, largely around his major assumptions regarding the nature of man. Not surprisingly, as was noted earlier, he sees man as naturally combative and competitive. And the compelling concern of that conflict, or competition, is power, or powers. But also, deeply influenced by new physical theories of Galileo and others, Hobbes' interpretations were based on the proposition that only matter and motion exist — bodies in motion.

In taking man, or the human body, as his subject, Hobbes pursues his subject from several perspectives — physiology, psychology, speech or language, morality, religion, politics, values. His grand plan was to develop a systematic, comprehensive, scientific philosophy which would explain all human behavior, all human activities, all facets of life. His goal was a conceptual scheme, all inclusive, linking all of these activities, all of these motions — from physiology, to psychology, to politics — all newly emerging sciences. Considering man as the primary kind of body in motion, Hobbes set out to explain his sensations, his desires, and behavior as results of simple internal motions, which resulted in complex external motions — a continual process until death.

What follows is not an analysis, not a reinterpretation, not a critique, not a summary of Hobbes' philosophy. It may be an over-simplification of the richness of his thinking. However, it seems worthwhile to call attention to selected portions of his writings in moving toward our subject — war. And finally, of course, toward the principal speaker on the subject, Nietzsche.

We are relatively certain that Nietzsche was familiar with Hobbes' writings. However, typically for Nietzsche, he did not engage the earlier writer in an explicit philosophical debate. What appears remarkable is the undeniable, unmistakable, intellectual genealogy linking the two thinkers. Both were brilliant, ruthless, dangerous and endangered, revolutionary, with unsurpassed intellectual vitality lasting a lifetime — one long, one short. Both were philosophers of nature, embracing the new sciences and scientific method as each understood it. Both held firmly to the belief that knowledge is power. Both lived in times of social and intellectual change, especially finding religion and religious beliefs and practices as demanding critical attention.

More specifically, as one would expect, Hobbes and Nietzsche shared distinctive similarities and distinctive differences. Separated by more than two hundred years, both were looking for a single principle by which to explain all human activities. Both focused on an old, but recently revived, notion — motion, the process of bodies in motion, of human bodies in motion, what Nietzsche would refer to as the process of "becoming". This was a defining concept for both, as was power.

Human bodies, for both Hobbes and Nietzsche, meant that human physiology was the place to start. Being philosophers, both were, or became, also psychologists. Both could be called scientific philosophers, as well as physiological psychologists. The two sciences — physiology and psychology — were directly tied to the nature of man. The science of politics, of paramount concern in Hobbes' philosophy, appears to have been of little direct concern for Nietzsche.

Hobbes' method of thinking scientifically did not include the important element of experiment, as it did later for Nietzsche. Experience was basic for Hobbes, but not in modern scientific procedure, not experimentation. Hobbes' analyses and interpretations preceded the later demands, especially by Nietzsche, that philosophy must become historical. Hobbes developed a tightly constructed systematic philosophy. Nietzsche was the antithesis of Hobbes — a rejection of system.

Hobbes and Nietzsche shared not only the basic ideas — bodies, motion, and power — but also the idea of values, particularly moral values, and the subjectivity and relativity of those values. Both had significant views on the value, or values, of persons, which meant human bodies. Both were convinced of the over-arching significance of words — of the use and function of language. Hobbes was an early advocate for clarity and precision in names, in definitions, and he shaped his analyses with the intention and expectation of revealing his "truths" with accuracy. Nietzsche extended the possibilities of the use of words — to speak philosophy — far beyond any previous thinker. Not immediate clarity, but obscurity was frequently his goal. He said of himself that he sought to be understood — but at a later time and with difficulty. Both philosophers, taking very innovative perspectives within the sciences of physiology and psychology, were aware of the dangerous nature of their interpretations — Nietzsche perhaps more even than Hobbes.

These then were the ideas and ways of engaging these ideas that Hobbes brought together in his systematic and coherent philosophy. Some of these he could trace back to early Greek philosophy — the study and reflection on *nature* and the *nature of man*; *bodies in motion*; *contrariety* and *conflict*; *religion* and *politics*. The more recent and new ideas which would radically revise our ways of thinking were *power* (particularly as interpreted by Machiavelli); *values*, especially moral values; and *language*, or speech. And, of course, the new access to these ideas being developed by the emerging sciences — *physiology* and *psychology*.

For the present purpose what follows are selected relevant words of Hobbes, in the hope of highlighting, if oversimplifying, a few of his original insights regarding these ideas. All of them are from Hobbes' *Leviathan*, Edited by Michael Oakeshott. First these:

> There be in animals, two sorts of *motions* peculiar to them: one called *vital*; begun in generation, and continued without interruption through their whole life; such as are the *course* of the blood, the *pulse*, the *breathing*, the *concoction, nutrition, excretion*, &c., to which motions there needs no help of imagination: the other is *animal motion*, otherwise called *voluntary motion*; as to *go*, to *speak*, to *move* any of our limbs, in such manner as is first fancied in our minds. That sense is motion in the organs and interior parts of man's body, caused by the action of the things we see, hear, &c.; and that fancy is but the relics of the same motion, remaining after sense, has been already said in the first and second chapters. And because *going, speaking*, and the like voluntary motions, depend always upon a precedent thought of *whither, which way*, and *what*; it is evident, that the imagination is the first internal beginning of all voluntary motion. And although unstudied men do not conceive any motion at all to be there, where the thing moved is invisible; or the space it is moved in is, for the shortness of it, insensible; yet that doth not hinder, but that such motions are. For let a space be never so little, that which is moved over a greater space, whereof that little one is part, must first be moved over that. These small beginnings of motion, within the body of man, before they appear in walking, speaking, striking, and other visible actions, are commonly called ENDEAVOUR. . . . This endeavour, when it is toward something which causes it, is called APPETITE, or DESIRE; the latter being the general name; and the other oftentimes restrained to signify the desire of food, namely *hunger* and *thirst*. And when the endeavour is fromward something, it is generally called AVERSION. These words, *appetite* and *aversion*, we have from the Latins; and they both of them signify the motions, one of approaching, the other of retiring. . . That which men desire, they are also said to LOVE: and to HATE those things for which they have aversion. So that desire and love are the same thing; save that by desire, we always signify the absence of the object; by love most commonly the presence of the same. So also by aversion, we signify the absence; and by hate, the presence of the object. . . Of appetites and aversions, some are born with men; as appetite of food, appetite of excretion, and exoneration, which may also and more properly be called aversions, from somewhat they feel in

their bodies; and some other appetites, not many. The rest, which are appetites of particular things, proceed from experience, and trial of their effects upon themselves or other men. For of things we know not at all, or believe not to be, we can have no further desire, than to taste and try. But aversion we have of things, not only which we know have hurt us, but also that we do not know whether they will hurt us or not. Those things which we neither desire, nor hate, we are said to *contemn*; CONTEMPT being nothing else but an immobility, or contumacy of the heart, in resisting the action of certain things; and proceeding from that the heart is already moved otherwise, by other more potent objects; or from want of experience of them. And because the constitution of a man's body is in continual mutation, it is impossible that all the same things should always cause in him the same appetites, and aversions: much less can all men consent, in the desire of almost any one and the same object. . . . But whatsoever is the object of any man's appetite or desire, that is it which he for his part calleth *good*: and the object of his hate and aversion, *evil*; and of his contempt, *vile* and *inconsiderable*. For these words of good, evil, and contemptible, are ever used with relation to the person that useth them: there being nothing simply and absolutely so; nor any common rule of good and evil, to be taken from the nature of the objects themselves; but from the person of the man, where there is no commonwealth; or, in a commonwealth, from the person that representeth it; or from an arbitrator or judge, whom men disagreeing shall by consent set up, and make his sentence the rule thereof (*L*, 47, 48, 49).

From these motions, which Hobbes calls "simple passions," — appetite, desire, love, aversion, hate, joy, and grief — he derives an extensive list of other complex passions and psychological states. For example, he defines fear as "aversion with opinion of hurt from the object." Or, hope as "appetite with an opinion of attaining." And all of these are built from the original small beginnings of voluntary motion , the *contraries* appetite and aversion, which he says are experienced as bodily, or physiological, motions, and which begin with imagination.

In the Introduction to the *Leviathan*, Hobbes says that the passions, which he later interprets in clear definitions, are alike in all men:

I say the similitude of *passions*, which are the same in all men, *desire, fear, hope*, &c.; not the similitude of the *objects* of the passions, which are the things *desired, feared, hoped*, &c.; for these the constitution individual, and particular education, do so vary, and they are so easy to be kept from our knowledge, that the characters of man's heart, blotted and confounded as they are with dissembling, lying, counterfeiting, and erroneous doctrines, are legible only to him that searcheth hearts. And though by men's actions

we do discover their design sometimes; yet to do it without com-
paring them with our own, and distinguishing all circumstances,
by which the case may come to be altered, is to decipher without
a key, and to be for the most part deceived, by too much trust, or
by too much diffidence; as he that reads, is himself a good or evil
man (L, 20).

Typical of Hobbes' insistence on the necessity of clear and simple
definitions in the processes of all thinking, writing, and speaking, he
begins his investigation of *power* as follows:

The power *of a man*, to take it universally, is his present means, to
obtain some future apparent good; and is either *original* or *instrumental*. *Natural power*, is the eminence of the faculties of body, or
mind: as extraordinary strength, form, prudence, arts, eloquence,
liberality, nobility. *Instrumental* are those powers, which acquired
by these, or by fortune, are means and instruments to acquire
more: as riches, reputation, friends, and the secret working of
God, which men call luck. For the nature of power, is in this
point, like to fame, increasing as it proceeds; or like the motion
of heavy bodies, which the further they go, make still the more
haste. The greatest of human powers, is that which is compounded of the powers of most men, united by consent, in one person,
natural, or civil, that has the use of all their powers depending on
his will; such as is the power of a commonwealth: or depending
on the wills of each particular; such as is the power of a faction or
of divers factions leagued. Therefore to have servants, is power;
to have friends, is power: for they are strengths united (L, 72).

In a process similar to that by which he derives the various passions,
Hobbes expands at length on the meaning of these different powers.
From the specification of what he holds to be the primary powers of a
man, he segues to the question of the *value* of a man:

The *value*, or WORTH of a man, is as of all other things, his price;
that is to say, so much as would be given for the use of his power:
and therefore is not absolute; but a thing dependant on the need
and judgment of another. An able conductor of soldiers, is of great
price in time of war present, or imminent; but in peace not so. A
learned and uncorrupt judge, is much worth in time of peace; but
not so much in war. As in other things, so in men, not the seller,
but the buyer determines the price. For let a man, as most men do
rate themselves at the highest value they can; yet their true value
is no more than it is esteemed by others. The manifestation of the
value we set on one another, is that which is commonly called
honouring, and dishonouring. To value a man at a high rate, is to
honour him; at a low rate, is to *dishonour* him (L, 73).

A little later Hobbes continues:

> WORTHINESS is a thing different from the worth, or value of a
> man; and also from his merit, or desert, and consisteth in a par-
> ticular power, or ability for that, whereof he is said to be worthy:
> which particular ability, is usually named FITNESS, or *aptitude*.
> For he is worthiest to be a commander, to be a judge, or to have
> any other charge, that is best fitted, with the qualities required
> to the well discharging of it; and worthiest of riches, that has the
> qualities most requisite for the well using of them: any of which
> qualities being absent, one may nevertheless be a worthy man,
> and valuable for something else. Again, a man may be worthy of
> riches, office, and employment, that nevertheless, can plead no
> right to have it before another; and therefore cannot be said to
> merit or deserve it. For merit presupposeth a right, and that the
> thing deserved is due by promise: of which I shall say more here-
> after, when I shall speak of contracts (*L*, 79).

When Nietzsche takes the question of the value of a person, or persons, and the relation of such value, or lack of it, to power or powers,
this becomes the basis of his famous challenge for mankind — the "revaluation of all values". His own interpretation of *values* will probably
be the entrance to his possible thinking regarding *war*. And, I believe
that he probably owed a debt to Hobbes for these earlier views.

But let's continue with Hobbes and his speaking about desires:

> By manners, I mean not here, decency of behaviour; as how one
> should salute another, or how a man should wash his mouth, or
> pick his teeth before company, and such other points of the *small
> morals*; but those qualities of mankind, that concern their living
> together in peace, and unity. To which end we are to consider,
> that the felicity of this life, consisteth not in the repose of a mind
> satisfied. For there is no such *finis ultimus*, utmost aim, nor *summum
> bonum*, greatest good, as is spoken of in the books of the old moral
> philosophers. Nor can a man any more live, whose desires are at
> an end, than he, whose senses and imaginations are at a stand.
> Felicity is a continual progress of the desire, from one object to
> another; the attaining of the former, being still but the way to
> the latter. . . .So that in the first place, I put for a general inclina-
> tion of all mankind, a perpetual and restless desire of power after
> power, that ceaseth only in death. And the cause of this, is not
> always that a man hopes for a more intensive delight, than he has
> already attained to; or that he cannot be content with a moder-
> ate power; but because he cannot assure the power and means
> to live well, which he hath present, without the acquisition of
> more. And from hence it is, that kings, whose power is greatest,
> turn their endeavors to the assuring it at home by laws, or abroad
> by wars: and when that is done, there succeedeth a new desire; in
> some, of fame from new conquests; in others, of ease and sensual
> pleasure; in others, of admiration, or being flattered for excel-

lence in some art, or other ability of the mind. . . . Competition
of riches, honour, command, or other power, inclineth to contention, enmity, and war: because the way of one competitor, to the
attaining of his desire, is to kill, subdue, supplant, or repel the
other. Particularly, competition of praise, inclineth to a reverence
of antiquity. For men contend with the living, not with the dead;
to these ascribing more than due, that they may obscure the glory
of the other. (*L*, 80, 81)

Hobbes' idea of the natural competitive nature of man is followed
by his famous interpretation of the so-called "state of nature" — the
natural condition of man prior to anything "artificial". It is a philosophical postulate used by Hobbes to further his account for the emergence
of the political state. He writes:

Nature hath made men so equal, in the faculties of the body, and
mind; as that though there be found one man sometimes mani-
festly stronger in body, or of quicker mind than another; yet when
all is reckoned together, the difference between man, and man, is
not so considerable as that one man can thereupon claim to himself any benefit, to which another may not pretend, as well as he.
For as to the strength of body, the weakest has strength enough
to kill the strongest, either by secret machination, or by confed-
eracy with others, that are in the same danger with himself. And
as to the faculties of the mind, setting aside the arts grounded
upon words, and especially that skill of proceeding upon general,
and infallible rules, called science; which very few have, and but
in few things; as being not a native faculty, born with us; nor attained, as prudence, while we look after somewhat else, I find
yet a greater equality amongst men, than that of strength. For
prudence, is but experience; which equal time, equally bestows
on all men, in those things they equally apply themselves unto
(*L*, 98).

Hobbes continues by analyzing the causes of quarrels among men,
and then writes this:

Hereby it is manifest, that during the time men live without a
common power to keep them all in awe, they are in that condition
which is called war; and such a war, as is of every man, against
every man. For WAR, consisteth not in battle only, or the act of
fighting; but in a tract of time, wherein the will to contend by
battle is sufficiently known: and therefore the notion of *time*, is to
be considered in the nature of war; as it is in the nature of weather. For as the nature of foul weather, lieth not in a shower or two
of rain; but in an inclination thereto of many days together; so the
nature of war, consisteth not in actual fighting; but in the known
disposition thereto, during all the time there is no assurance to
the contrary. All other time is PEACE. . . .Whatsoever therefore

is consequent to a time of war, where every man is enemy to every man; the same is consequent to the time, wherein men live without other security, than what their own strength, and their own invention shall furnish them withal. In such condition, there is no place for industry; because the fruit thereof is uncertain; and consequently no culture of the earth; no navigation, nor use of the commodities that may be imported by sea; no commodious building; no instruments of moving, and removing, such things as require much force; no knowledge of the face of the earth; no account of time; no arts; no letters; no society; and which is worst of all, continual fear, and danger of violent death; and the life of man, solitary, poor, nasty, brutish, and short (*L*, 100).

Investigating and evaluating religion within the context of interpreting power and war was important for Machiavelli, for Hobbes, and as we will see later, important for Nietzsche. Both Machiavelli and Hobbes had included the passion of fear in their discussions. Hobbes gave this succinct reading:

Fear of power invisible, feigned by the mind, or imagined from tales publicly allowed, RELIGION; not allowed, SUPERSTITION. And when the power imagined, is truly such as we imagine, TRUE RELIGION (*L*, 51).

During the period when Hobbes was living and writing, religious criticism and controversy were climbing to new heights. At the center of the storm, so involved was Hobbes that he devoted the third and fourth parts of *Leviathan* to seeking to identify the place of religion in his design for a new society.

Finally, philosophical analysis of language in Hobbes' thinking also reached new dimensions and significance. Undoubtedly, it was of no little importance that he was roughly contemporary with two Englishmen of similar persuasion and genius — William Shakespeare and Samuel Johnson. Nietzsche, influenced by Shakespeare and by Hobbes, was overwhelmed by words and the power of words, in the use of which he became a master. From Hobbes, these several words seem important:

The general use of speech, is to transfer our mental discourse, into verbal; or the train of our thoughts, into a train of words; and that for two commodities, whereof one is the registering of the consequences of our thoughts; which being apt to slip out of our memory, and put us to a new labour, may again be recalled, by such words as they were marked by. So that the first use of names is to serve for *marks*, or *notes* of remembrance. Another is, when many use the same words, to signify, by their connexion and order, one to another, what they conceive, or think of each matter; and also what they desire, fear, or have any other passion for. And for this use they are called *signs*. Special uses of speech are these; first, to register, what by cogitation, we find to be the cause of any thing, present or past; and what we find things present or past may produce, or effect; which in sum, is acquiring of arts. Secondly, to show to others that knowledge which we have attained, which is, to counsel and teach one another. Thirdly, to make known to others our wills and purposes, that we may have the mutual help of one another. Fourthly, to please and delight ourselves and others, by playing with our words, for pleasure or ornament, innocently. . . .To these uses there are also four correspondent abuses. First, when men register their thoughts wrong, by the inconstancy of the signification of their words; by which they register for their conception, that which they never conceived, and so deceive themselves. Secondly, when they use words metaphorically; that is, in other sense than that they are ordained for; and thereby deceive others. Thirdly, by words, when they declare that to be their will, which is not. Fourthly, when they use them to grieve one another; for seeing nature hath armed living creatures, some with teeth, some with horns, and some with hands, to grieve an enemy, it is but an abuse of speech, to grieve him with the tongue, unless it be one whom we are obliged to govern; and then it is not to grieve, but to correct and amend. . . .Seeing then that truth consisteth in the right ordering of names in our affirmations, a man that seeketh precise truth had need to remember what every name he uses stands for, and to place it accordingly, or else he will find himself entangled in words, as a bird in lime twigs, the more he struggles the more belimed. And therefore in geometry, which is the only science that it hath pleased God hitherto to bestow on mankind, men begin at settling the significations of their words; which settling of significations they call *definitions*, and place them in the beginning of their reckoning. By this it appears how necessary it is for any man that aspires to true knowledge, to examine the definitions of former authors; and either to correct them, where they are negligently set down, or to make them himself. For the errors of definitions multiply themselves according as the reckoning proceeds, and lead men into absurdities, which at last they see, but cannot avoid, without reckoning anew from the beginning, in which lies the foundation of their errors. . . .So that in the right definition of names lies the first use of speech; which is the acquisition of science: and in wrong, or no definitions, lies the first abuse; from which proceed all false and senseless tenets; which make those men that take their instruction from the authority of books, and not from their own meditation, to be as much below the condition of ignorant men, as men endued with true science are above it. For between true science and erroneous doctrines, ignorance is in the middle. Natural sense and imagination are not subject to absurdity. Nature itself cannot err; and as men bound in copiousness of language, so they become more wise, or more mad than ordinary. Nor is it possible without letters for any man to become either excellently wise, or, unless his memory be hurt by disease or ill constitution of organs, excellently foolish. For words are wise men's counters, they do but reckon by them; but they are the money of fools, that value them by the authority of an Aristotle, a Cicero, a Thomas, or any other doctor whatsoever, if but a man (*L*, 34, 36, 37).

C. B. Macpherson begins the Introduction to his edition of *Leviathan* with these comments:

Why, in the second half of the twentieth century, do we still read Hobbes, who wrote three centuries ago? Why the perennial fascination of his works? Or perhaps we should be asking, why the recurring fascination of his works, for it is in our own time that a heightened interest in them has developed. What has happened to twentieth-century man to bring him closer to Hobbes than were the men of the intervening centuries? When the question is put in this way some answers suggest themselves at once. Our world is obsessed with problems of power, and Hobbes was an analyst of power. We want not only to understand and to control power, but also to harness it to right. So did Hobbes. He exposed the lineaments of power more clearly than anyone had done since Machiavelli, more systematically than anyone had ever done, and than most have done since.

So we may say that the twentieth century has brought us closer to an appreciation of Hobbes on three counts: power, peace, and science. Our century has compelled us to new interest in his subject matter — the power relations, necessary, possible, and desirable, between men; in his purpose — to find a way to peace and "commodious living"; and in his method — the method of science (*LE*, 9, 10, 11).

For now, I rest my case for having Hobbes called to address some of the important issues and perspectives that might help in getting deeper into understanding these "calamities". (Remember, Hobbes had said, "Now all such calamities as may be avoided by human industry arise from war, . . .") After Hesiod, Heraclitus, and Hobbes, the fourth, and final witness, is a person who, in my view, appears to be little known or recognized in Nietzschean literature, but whose stimulating and controversial ideas probably had as great an influence as any single person on Nietzsche's thinking. No one contributed more, I believe, if indirectly, to his possible interpretation of the subject under scrutiny — *war*. That person was Johann Jakob Bachofen.

J. J. Bachofen was a Swiss jurist, social philosopher, cultural historian, and a contemporary of Hegel, Marx, Schopenhauer, and a younger Nietzsche. He had a great deal to say and to offer to Nietzsche. Interestingly and uncannily, Hobbes, Bachofen, and Nietzsche shared similar situations and characteristics. All three philosophers were brilliant, creative, innovative, comprehensive, controversial, revolutionary, dangerous. All three were interested in philosophy of nature. All were welcoming enthusiastically the opportunities offered by the new sciences and scientific method. And all were gathering together — each according to his own interests and aims — essentially most of the same foundational ideas. And what were these? Power, contraries, conflict, bodies, values, scientific method, religion.

However, Bachofen would propose two revolutionary ideas. First, he would apply the methods of science to history. Whereas Hobbes believed that politics could, and should, become the science of politics, Bachofen believed similarly regarding human history. We will look at his innovative and radical approach. Second, in investigating and interpreting the process of history, he naturally encountered, and emphasized, what had clearly long been a subject in art and literature — the human body in two forms, female bodies and male bodies — *sexuality*. Studying history, Bachofen reinserted the dominant place and power of sexuality. And not only in history, but also into philosophy. Nietzsche later, almost certainly influenced by Bachofen in major ways, would write, ". . . and often I have asked myself whether, taking a large view, philosophy has not been merely an interpretation of the body and a *misunderstanding of the body*."

The implications of this simple reminder — that human bodies in two forms, female and male, make up the natural world, necessarily separate, different, but interdependent in a unity — were enormous. Perhaps what was entailed was a reconsideration, a reinterpretation of power or powers, of conflict or war, of contraries, of values, of experience, of speech, of physiology, of psychology, of religion. This was, and would continue to be transformative in understanding humans — both in nature and in culture. Sexuality, the basic formative pattern of opposites — male and female — became the basis of Nietzsche's devastating criticism of Christianity and Christian morality. There is a strong likelihood, an imperative, that his possible critique of war would be similarly based.

As to Bachofen, it appears that his work became available and recognized by English readers only as late as 1967, with the publication of a collection of his writings, *Myth, Religion, and Mother Right*, translated by Ralph Manheim. And perhaps it was in this publication that the important link between Bachofen and Nietzsche began to emerge as a titillating opportunity for further attempts at solving the puzzle that was Nietzsche. (Such was my own discovery.)

During Bachofen's lifetime (1815–1887) there was a continuing and growing interest in the philosophical community in opening many new accesses to the study and understanding of the natural world — of nature. And science was the catalyst. After centuries of theological and philosophical dogmatizing, nature, the earth, life, the human body, anything related to, or part of, nature — all of these became the focus of many new sciences — biology, physiology, psychology, medicine, physics, archeology, anthropology, ethics, aesthetics, and of course, history. We have seen how physiology, psychology, and politics in Hobbes' philosophy became his major sciences of interest. Also, it was clear that in the seventeenth century, as a new "scientific philosopher", Hobbes found it intellectually and politically expedient and prudent to devote a large part of his major work to interpreting and justifying theology and religion.

When Hobbes made his famous claim of there having been a situation of human community which he called the "state of nature," he was offering a philosophical postulate from which he wanted to derive his theory of politics. He did the same when he described man's situation in the so-called "state of nature" as being "a war of every man against every man."

Bachofen, writing two centuries after Hobbes, as one would expect, was much more imbued with the importance of an empirical approach, or method of science. When he takes history as the focus of his investigations, to apply the scientific method to historical events and observations, he would claim the discovery of an early stage of human development — not as a philosophical presupposition, but based on an interpretation of available evidence. Whereas Hobbes saw the necessary connection between nature and culture as a logical process, Bachofen saw the connection as a process of evolution, a process of development occurring in stages. These two structured their theories around these same ideas — bodies, motion, power, values, conflict, contraries, the power of religion. The addition of scientific history and sexuality by Bachofen to the intellectual dynamics of the latter part of the nineteenth century cannot be overestimated. The threat that his thinking posed to traditional and conventional thinking was equal to that of Hobbes before him. Both contributed greatly to the eruption that would result with Nietzsche's dangerous mission.

In 1967, a book entitled *Myth, Religion, and Mother Right*, Selected Writings of J. J. Bachofen, was published by Princeton University Press, with Preface by George Boas and Introduction by Joseph Campbell. In order to establish and support what appears to be an almost undeniable strong influence of Bachofen's thinking and writings on Nietzsche, what follows are several quotations from both Boas and Campbell, as well as direct quotations from the author, Bachofen. Boas begins the Preface:

> The name of Johann Jakob Bachofen, if mentioned at all in books of reference is attached to a theory of social development which maintains that the first period of human history was matriarchal. And if any discussion of the theory is added, it will be to the effect that it is almost universally discredited. As a matter of fact this is only a small part of Bachofen's contribution to social philosophy, and it would be perhaps more appropriate, if labels are required, to list him among either ethnologists or sociologists. For, as the studies in this volume will show, his attitude toward cultural history was not that of the empirical anthropologist or that of the annalist. His focus of interest was the inner life of human beings rather than what he called the externals of human development. He was more concerned with literature, language, architecture, and the other arts than with economic factors, military adventures, territorial expansion, the succession of rulers, population growth, and revolutions, whether in isolation from one another or all in a grand hodge-podge (MR, xi).

Campbell begins the Introduction:

> It is fitting that the works of Johann Jakob Bachofen (1815–1887) should have been rediscovered for our century, not by historians or anthropologists, but by a circle of creative artists, psycholo-

gists, and literary men.. . . For Bachofen has a great deal to say to artists, writers, searchers of the psyche, and, in fact, anyone aware of the enigmatic influence of symbols in the structuring and moving of lives: the lives of individuals, nations, and those larger constellations of destiny, the civilizations that have come and gone. . . (MR, xxv).

Further into the Introduction Campbell refers to Bachofen's "young friend Friedrich Nietzsche", and writes:

Nietzsche, who came to Basel in 1869 as a young professor of classical philology and for the next half decade was a frequent guest in Bachofen's home (spending Sundays, however, with his idol Richard Wagner in Lucerne, whose dates, 1813-1883, again approximately match Bachofen's span of years), saw the dialectic of history, and of individual biography as well, in terms of an un-relenting conflict between the forces of disease, weakness, and life-resentment, on the one hand, and on the other, the courage and determination to build a life forward toward a realization of potentials (MR, xlvi, xlvii).

A few paragraphs later Campbell writes:

The arrival in Basel, three years later, of Nietzsche, brought a new brilliance to the Bachofen domestic circle, and it was about that time that signs began to appear, as well, of a new, significant, and rapidly increasing scientific appreciation of his published works — not, indeed, from the classical circles of academic hardshelled crabs to which he had turned, at first, in vain; but from the un-foreseen quarters of a new science, anthropology (MR, li).

Further, with reference to the relationship between Bachofen and Nietzsche, Boas writes:

There are certain themes in Bachofen which have a strong similar-ity to those of Nietzsche. Though Bachofen's name does not ap-pear in any of the indexes to any of Nietzsche's works, Nietzsche was a great admirer of Bachofen's colleague, Jakob Burckhardt, and Burckhardt himself was an admirer of Bachofen. . . It is to be noted also that in his early *Birth of Tragedy* Nietzsche took over Bachofen's terms the Dionysiac and the Apollonian, for two types of will, the creative and the contemplative, and that he also main-tained that they were fused into one in the Greek tragedy before the time of Euripides (MR, xx).

Later, significant differences appear between Nietzsche's philoso-phy and that of Bachofen; however, one of the important similarities is that of the role assigned by both to symbols and myths. Bachofen him-self speaks of "Dionysian truth" and "Dionysian religion". Even though, as Boas points out, Bachofen's name apparently does not appear in any of the indexes of Nietzsche's works, and although apparently no translator, biographer, or critic of Nietzsche's works has referred to his lengthy acquaintance with Bachofen, it is reasonable to assume that the young scholar would have had many absorbing and invigorating conversations with the older scholar. And equally reasonable to consider what possible influences Bachofen's thinking may have had on this young classical philologist. In a short biographical sketch written by Bachofen in 1854, he wrote:

> I was drawn to the study of law by philology. It is here that I started and hither that my legal studies led me back. In this respect my attitude toward the field has always remained unchanged. Roman law has always struck me as a branch of classical and particularly of Latin philology, hence as part of a vast field encompassing the whole of classical antiquity. . . . [I]t was ancient and not modern Roman law that I really wanted to study (MR, 3).

Bachofen's interest was Latin philology, Nietzsche's interest was mainly Greek philology. As a jurist, Bachofen entered into the study of the history of Roman law. At the age of twenty-four, he began a series of trips, spending time in universities and museums in Paris and London. Of his stay in London he wrote, "I was fascinated by the life of the law courts with all their patriarchal pomp, but there was also the British Museum with its treasures." Bachofen's literary studies led him, he writes, to "Winckelmann's History of Art." He continues:

> But to my reading of Winckelmann's works I owe an enjoyment of a far higher order — indeed, one of the greatest pleasures of my whole life. Since then I have dwelt much in the regions that it opened, especially at times when everything else seemed to lose interest for me. Ancient art draws our heart to classical antiquity, and jurisprudence our mind. Only together do the two confer a harmonious enjoyment and satisfy both halves of man's spirit. Philology without concern for the works of art remains a lifeless skeleton. . . In my wanderings through the museums of Italy my attention was soon attracted to one aspect of all their vast treasures, namely mortuary art, a field in which antiquity shows us some of its greatest beauties. When I consider the profound feeling, the human warmth that distinguishes this realm of ancient life, I am ashamed of the poverty and barrenness of the modern world. The ancient tombs have given us a well-nigh inexhaustible wealth. At first we may regard the study of tombs as a specialized field of archeology, but ultimately we find ourselves in the midst of a truly universal [religious] doctrine. All the treasures that fill our museums of ancient art were taken from tombs. . . Antiquity

made full use of the symbolic, most enduringly and profoundly in its art. . . The sun warms and illuminates these resting places of the dead so wonderfully and infuses the abodes of horror with the magic of joyous life. What beauty there must have been in an age whose very tombs can still arouse so much yearning for it! What a vast abundance of beautiful ethical ideas the ancients drew from their myths. The treasure house that encompasses their oldest memories of history serves also as a source of the oldest ethical truths, and provides consolation and hope for the dying (MR, 10, 11, 12, 13).

More study, more reading, more trips to Italy and London and the British Museum, and he wrote:

Rome became the fulfillment and end of a cultural era spanning a millennium. . . . I found myself in a period of transition such as occurs in the life of every striving man. . . . Ever since then my guiding thought has been the religious foundation of all ancient thinking and life. . . . What I am engaged upon now is the true study of nature. The material alone is my preceptor. It must first be assembled, then observed, and analyzed (MR, 14, 15, 16).

Bachofen's learning, interests, and experience — jurisprudence, history of Roman law, classical philology, symbolism and mythology — coalesced into what may be called "cultural history". He was interested not only in individual symbols or myths, or individual cultures, but in discovering what he called "the universal law of history" — in developing his theory of "scientific history". The development of his thinking reflected, and was consistent with, theoretical historical thinking of his time. The dominant ideas regarding history were first — history develops in discernible stages. Not to be understood as cause and effect, but rather as one stage containing the potential, or seeds, for the succeeding stage, with each stage taken up into the next stage. Second — that this development was unilinear, involving a series of changes, usually from primitive to more advanced, from simple or lower to more complex or higher, an ascending process. Third — that there were so-called "law" or "laws" which should be discoverable. Perhaps Bachofen might have referred to this "universal law of history" as a pattern, or patterns — as natural or chance configuration, repeatable or recurring again and again.

Later Nietzsche would reject the notion of a "law" or "laws" of nature, preferring instead the idea of "necessity" or "necessities", repeatable again and again, or "recurring eternally". In any case, in considering the evidence, the material to be "assembled, observed and analyzed" consisted of the symbols in art and the myths in literature. For symbols and myths, and their interpretations, were the source of understanding the meaning of life for the ancients. This meant a study of nature and what he referred to as "nature religions". Bachofen was persuaded that he, and we, needed and were able to have access to, pre-classical and pre-Christian culture and the religion. He wanted to study and learn from myths and symbols, but also from all available sources in classical literature, language, and the arts. To effectively develop his theory of the history of culture, he needed to be able to trace the development, the transformations, from these pre-classical, ancient cultures through various stages up to, and including, the present — the Christian era.

Before hearing further some of Bachofen's words, this may be noted. He perceived the fundamental duality of nature, and of life, as residing primarily in the oppositions of male and female, and life and death, and in the constant conflict between these opposites. Sexuality, power or powers, both natural and "supernatural", values, rankings, attitudes, beliefs, behaviors. And he interpreted the highest force shaping cultures, the basic foundation of any culture, as being religion.

Bachofen's theory was one interpreting history as a linear progression from lower forms — which he calls "matriarchal" or "maternal" — to higher forms — which he calls "patriarchal" or "paternal". The most important principle operating as stimulus, or catalyst, for the emergence of a new and different stage was the intensification, abuse, or perversion of *power* by either the female/woman/mother or the male/man/father. Boas, in his Preface to Bachofen's writings reminds us, "Bachofen's theory of a matriarchal society out of which modern patriarchal societies evolved was accepted pretty generally among sociologists until about the beginning of the twentieth century. It was the classic pattern for historians to follow." As earlier in his life, once again however, Bachofen's interpretations were derided, dismissed, and virtually disappeared.

The details of his theory is the substance of his book, *Myth, Religion, and Mother Right*. Here are some of his words which serve mostly to indicate the basic themes of his attempt to describe five separate stages in this evolutionary process. Bachofen's own Introduction begins:

The present work deals with a historical phenomenon which few have observed and no one has investigated in its full scope. Up until now archeologists have had nothing to say of mother right. The term is new and the family situation it designates unknown. The subject is extremely attractive, but it also raises great difficulties. The most elementary spadework remains to be done, for the cultural period to which mother right pertains has never been seriously studied. Thus we are entering upon virgin territory.

We find ourselves carried back to times antedating classical antiquity, to an older world of ideas totally different from those with which we are familiar. . . The matriarchal organization of the family seems strange in the light not only of modern but also of classical ideas. . . The main purpose of the following pages is to set forth the moving principle of the matriarchal age, and to give its proper place in relationship both to the lower stages of development and to the higher levels of culture. . . In this way I hope to restore the picture of a cultural stage which was overlaid or totally destroyed by the later development of the ancient world. . . Polybius' message about the matriarchal genealogy of the hundred noble families among the Epizephyrian Locrians suggests two further observations which have been confirmed in the course of our investigations: (1) mother right belongs to a cultural period preceeding that of the patriarchal system; (2) it began to decline only with the victorious development of the paternal system.

The principles which we have deduced from a few observations are confirmed in the course of our investigation by an abundance of data. . . The prestige of womanhood among these people was a source of astonishment to the ancients, and gives them all, regardless of individual coloration, a character of archaic sublimity that stands in striking contrast to Hellenic culture (*MR*, 70, 71).

A little later Bachofen writes:

Both the mythical and the strictly historical traditions present very similar pictures of the system. Products of archaic and of much later periods show such an astonishing accord that we almost forget the long interval between the times when they originated. . . Precisely in regard to the most important aspect of history, namely, the knowledge of ancient ideas and institutions, the already shaky distinction between historic and prehistoric times loses its last shred of justification (*MR*, 73).

Bachofen traces what he calls the "juridical aspect" of mother right, indicating that in matriarchal cultures the sister, daughter, and youngest child are found to have the favored position over the brother, son, or oldest child. Considering what he calls the "ethical aspect", he writes:

The ethical aspect strikes a resonance in a natural sentiment which is alien to no age: we understand it almost spontaneously. At the lowest, darkest stages of human existence the love between the mother and her offspring is the bright spot in life, the only light in the moral darkness, the only joy amid profound misery. By recalling this fact to our attention, the observations of still living peoples of other continents has clarified the mythical tradition which represents the appearance of the (father lovers) as an important turning point in the development of human culture. The close relation between child and father, the son's self-sacrifice for his begetter, requires a far higher degree of moral development than mother love, that mysterious power which equally permeates all earthly creatures. Paternal love appears later. The relationship which stands at the origin of all culture, of every virtue, of every nobler aspect of existence, is that between mother and child; it operates in a world of violence as the divine principle of love, of union, of peace. Raising her young, the woman learns earlier than the man to extend her loving care beyond the limits of the ego to another creature and to direct whatever gift of invention she possesses to the preservation and improvement of this other's existence. Woman at this stage is the repository of all culture, of all benevolence, of all devotion, of all concern for the living and grief for the dead (*MR*, 79).

Next Bachofen addresses the controversial question of the religious aspect of matriarchy:

The religious foundation of matriarchy discloses this system in its noblest forms, links it with the highest aspects of life, and gives a profound insight into the dignity of that primordial era which Hellenism excelled only in outward radiance, not in depth and loftiness of conception. . . There is only one mighty lever of all civilizations, and that is religion. Every rise and every decline of human existence springs from a movement that originates in this supreme sphere. Without it no aspect of ancient life is intelligible, and earliest times in particular remain an impenetrable riddle. Wholly dominated by its faith, mankind in this stage links every form of existence, every historical tradition, to the basic religious idea, sees every event in a religious light, and identifies itself completely with its gods. If especially matriarchate must bear this hieratic imprint, it is because of the essential feminine nature, that profound sense of the divine presence which, merging with the feeling of love, lends woman, and particularly the mother, a religious devotion that is most active in the most barbarous times. The elevation of woman over man arouses our amazement most especially by its contradiction to the relation of physical strength. The law of nature confers the scepter of power on the stronger. If it is torn away from him by feebler hands, other aspects of human nature must have been at work, deeper powers must have made their influence felt. . . The religious primacy of motherhood leads to a primacy of the mortal woman; Demeter's exclusive bond with Kore leads to the no less exclusive relation between mother and daughter; and finally, the inner link between the mystery and the chthonian–feminine cults leads to the priesthood of the mother, who here achieves the highest degree of religious consecration.

These considerations bring new insight into the cultural stage characterized by matriarchy. We are faced with the essential greatness of the pre-Hellenic culture: in the Demetrian mystery and the religious and civil primacy of womanhood it possessed the seed of noble achievement which was suppressed and often destroyed by later developments. . . But mystery is the true essence of every religion, and wherever woman dominates religion or life, she will cultivate the mysterious. Mystery is rooted in her very nature, with its close alliance between the material and the supersensory; mystery springs from her kinship with material nature. . . Hellenism is hostile to such a world. The primacy of motherhood vanishes, and its consequences with it. The patriarchal development stresses a completely different aspect of human nature, which is reflected in entirely different social forms and ideas (*MR*, 84, 85, 87, 89).

Bachofen continues:

The central idea that I have emphasized from the outset, the relationship between the primacy of women and the pre-Hellenic culture and religion, is eminently confirmed by the very phenomena which, when viewed superficially and out of context, seem to argue most against it. Whenever the older mystery religion is preserved or revived, woman emerges from the obscurity and servitude to which she was condemned amid the splendor of Ionian Greece and restored to all her pristine dignity. . . No era has attached so much importance to outward form , to the sanctity of the body, and so little to the inner spiritual factor. . . and none has been so given to lyrical enthusiasm, this eminently feminine sentiment, rooted in the feeling of nature. In a word, matriarchal existence is regulated naturalism, its thinking is material, its development predominantly physical. Mother right is just as essential to this cultural stage as it is alien and unintelligible to the era of patriarchy (*MR*, 90, 92).

Having made his case for the significance of mythology and for the necessity of attempting to interpret and understand these earliest forms of human culture; and having devoted an enormous amount of research to understanding the inner structure of the matriarchal system, especially assigning to religion the dominant position in that system, Bachofen next turned his attention to history, to determine the relation of the matriarchal culture to other cultural stages. He writes:

We shall come face to face with a new aspect of history. We shall encounter great transformations and upheavals which will throw a new light on the vicissitudes of human destiny. Every change in the relation between the sexes is attended by bloody events; peaceful and gradual change is far less frequent than violent upheavals. Carried to the extreme, every principle leads to the victory of its opposite; even abuse becomes a lever of progress; supreme triumph is the beginning of defeat. Nowhere is man's tendency to exceed the measure, his inability to sustain an unnatural level, so evident; and nowhere is the scholar's capacity for entering into the wild grandeur of crude but gifted peoples, and for making himself at home among utterly strange ideas and social forms, put to so rigorous a test. . . But it is precisely the connection of the sexual relationship and the lower or higher interpretation with the totality of life and the destinies of nations that relates our study directly to the cardinal questions of history (MR, 92, 93, 99).

Bachofen's investigations led him to speak at length regarding what he called the "Dionysian religion." Here, briefly is one important excerpt:

The magic power with which the phallic lord of exuberant natural life revolutionized the world of women is manifested in phenomena which surpass the limits of our experience and our imagination. Yet to relegate them to the realm of poetic invention would betoken little knowledge of the dark depths of human nature and failure to understand the power of a religion that satisfied sensual as well as transcendent needs. It would mean to ignore the emotional character of woman, which so indissolubly combines immanent and transcendent elements, as well as the overpowering magic of nature in the luxuriant south.

Throughout its development the Dionysian cult preserved the character it had when it first entered into history. With its sensuality and emphasis on sexual love, it presented a marked affinity to the feminine nature, and its appeal was primarily to women; it was among women that it found its most loyal supporters, its most assiduous servants, and their enthusiasm was the foundation of its power. Dionysus is a woman's god in the truest sense of the word, the source of all woman's sensual and transcendent hopes, the center of her whole existence. It was to women that he was first revealed in his glory, and it was women who propagated his cult and brought about its triumph. A religion which based even its higher hopes on the fulfillment of the sexual commandment, which established the closest bond between the beatitude of supersensory existence and the satisfaction of the senses, could not fail, through the erotic tendency it introduced into the life of women, to undermine the Demetrian morality and ultimately to reduce the matriarchal existence to an Aphroditean hetaerism patterned after the full spontaneity of natural life. . . All life was molded by the same trend, as is shown above all by the ancient tombs which, in a moving paradox, became the chief source of our culture of the erotic sensuality of Dionysian womanhood. Once again we recognize the profound influence of religion on the development of all culture. The Dionysian cult brought antiquity the highest development of a thoroughly Aphroditean civilization, and lent it that radiance which overshadowed all the refinement and all the art of modern life (MR, 101, 102).

Later in the book he writes this:

The progress from the maternal to the paternal conception of man forms the most important turning point in the history of the relations between the sexes. The Demetrian and the Aphroditean-hetaeric stages both hold to the primacy of generative motherhood, and it is only the greater or lesser purity of its interpretation that distinguishes the two forms of existence. But with the transition to the paternal system occurs a change in fundamental principle; the older conception is wholly surpassed. An entirely new attitude makes itself felt. The mother's connection with the child is based on a material relationship, it is accessible to sense perception and remains always a natural truth. But the father as begetter presents an entirely different aspect. Standing in no visible relation to the child, he can never, even in the marital relation, cast off a certain fictive character. Belonging to the offspring only through the mediation of the mother, he always appears as the remote potency. As promoting cause, he discloses an immateriality over against which the sheltering and nourishing mother appears as (matter), as (place and house of generation), as (nurse) (MR, 109).

One last observation from Bachofen reads:

This homogeneity of matriarchal ideas is confirmed by the favoring of the left side over the right side. The left side belongs to the passive feminine principle, the right side to the active masculine principle. The role played by the left hand of Isis in matriarchal Egypt suffices to make the connection clear. But a multitude of additional data prove its importance, universality, primordiality, and freedom from the influence of philosophical speculation. Customs and practices of civil and religious life, peculiarities of clothing and headdress, and certain linguistic usages reveal the same idea. . . Another no less significant manifestation of the same basic law is the primacy of the night over the day which issued from its womb. . . Already the ancients identified the primacy of the night with that of the left, and both of these with the primacy of the mother. And here, too, age-old customs, the reckoning of time according to nights, the choice of the night as a time for battle, for taking counsel, for meting out justice, and for practicing the cult rites, show that we are not dealing with abstract philosophical ideas of later origin, but with the reality of an original mode of life. . . Extension of the same idea permits us to recognize the religious preference given to the moon over the sun, of the conceiving earth over the fecundating sea. . . . (MR, 77).

Nietzsche recognized the inexhaustible opportunities in symbolic opposites and seized these opportunities, both to re-explore traditional opposites and to create and experiment with new ones of his own — most in the service of speaking about female and male, woman and man. It was rapidly becoming clear to both Bachofen and Nietzsche that the prototype of all opposition or contrariety is the *contrariety of sexuality*. In describing the type of character which is his Zarathustra, Nietzsche writes:

He contradicts with every word, this most affirmative of all spirits, all are in him bound together in a new unity. . . It is precisely this compass of space, in this access to opposites that Zarathustra feels himself to be the *highest species of all existing things*. . . (EH, 106, 107).

Bachofen ends his Introduction with these words:

The present book makes no other claim than to provide the scholarly world with a new and well-nigh inexhaustible material for thought. If it has the power to stimulate, it will gladly content itself with the modest position of a preparatory work, and cheerfully accept the common fate of all first attempts, namely to be disparaged by posterity and judged only on the basis of its shortcomings (MR, 120).

It would be difficult to over-estimate the probable influences on Nietzsche's life and career from this long acquaintance between these two imaginative, creative, revolutionary thinkers. Both set out to radically change the human world by changing our understanding of that world. Bachofen, the elder, bequeathed to Nietzsche, the younger, in the latter's early, formative, and impressionable years, a treasure of energy, power, excitement, possibilities. Nietzsche's obvious natural talent for, and love of, ideas and words to express these ideas, and his awareness of the power of words, must have been stimulated beyond any anticipation. Words to learn more about the past, to understand the present, and to help shape the future — these were the aims of both. Bachofen's name may not appear in any of the indexes of Nietzsche's books, but his intellectual genes are evident in many places.

These are the themes in Bachofen's legacy that seem to stand out. First — the notion of a scientific approach to understanding history, particularly cultural history, with his focus on language, literature, architecture, and other arts. Second — the emphasis on the idea that the key element of cultural history was the study of nature, the continuing changes in the male–female relationship, between the feminine principle and the masculine principle, the maternal and the paternal, sexuality, the centrality and the sanctity of the body. Third — the interpretation of the power and force of religion as having first rank among the creative forces of culture, as the foundation of culture. Fourth — the necessity of recognizing that the language of myths and symbols was the key to understanding the earliest stages of the process of cultural history. Myths and symbols provided reliable evidence for early thinking regarding the natural primal relation, that of the sexes, and also regarding questions of life and death, of generating life and destroying life, and more.

The background, then, for attempting to tease from Nietzsche what his thoughts regarding war might have been, has been provided by these four — Hesiod, Heraclitus, Hobbes, and Bachofen. And together, in different centuries and different locations, from vastly different perspectives, they offer an astonishing array of essentially the same ideas, or at least closely interwoven ideas. These are: *conflict, contrariety, power, values, nature, culture, mythology, religion, sciences, history, physiology, psychology,* and *sex or sexuality.*

All of these ideas, or areas of interest, or sciences, I believe, form a pattern (however difficult it may become to interpret the pattern) from which a credible "Nietzschean critique of war" may emerge and become visible. Perhaps the preferable procedure is to attempt to address some of these ideas separately and then to indicate the mingling that frequently and necessarily occurs among them.

Nietzsche's perspectives on Western culture, and especially his own German culture — the evidence of the continuing process of corruption, decay, and disintegration — signified to him grave and present danger, a looming possible catastrophe. Early in his career, he was beginning to define what would become his life's task — to probe ever more deeply into the source, or sources, of this rapidly deteriorating condition. And he was becoming ever more aware of the increasing dangers to himself as he ventured further into what had been for centuries forbidden places. He eventually defined his task as that of exposing and ultimately destroying what he would uncover.

PART THREE

As a philosopher and as a classical philologist, Nietzsche would have read *about* war, conflict, strife. From the early Greek poets, historians, dramatists, philosophers, to his own age, the subject — treated directly or indirectly, literally or metaphorically — could have been perceived by Nietzsche as the dominating theme in literature. And without a doubt, his familiarity with the Old and New Testaments would have brought a resounding and agonizing repetition of the subject. He would have thought *about* conflict or war, written *about* it, spoken *about* it. But as with all of his thinking, it was his own experience, or experiences — internal and external — which ultimately provided the major sources for thinking and writing. Actually, it was his consciousness of these experiences and his constant demand for unrelenting honesty — the avoidance of self-deception and deception of others — that opened him to his unexpected and dangerous "truths".

The idea of *experience* appears to have increasingly captured the attention and imagination of thinkers and writers of different persuasions, none more than Nietzsche. And during his creative period, there was growing awareness that experiences, although always individual and unique, nevertheless took two similar but different forms — female and male. Also, that these differences in experiences would necessarily reflect the physiological differences in male and female bodies.

In addition to the idea of experience, another related major idea, with expanding present and possible future significance, was the idea of *consciousness*, that is, acute awareness of, interest in, concern with, especially one's own experience or experiences. This interest in consciousness quickly expanded into species consciousness, class consciousness, race consciousness, gender or sexual consciousness, national consciousness, and historical or time consciousness. This usually included the growing awareness of, interest in, concern with one's own biological, or social or economic *rank* in society, a feeling of identification with those belonging to the same group or class as oneself.

Charles Darwin's theory relating to species consciousness and Karl Marx's theory relating to class consciousness, along with the interest of both in historical consciousness — historical development or historical evolution — were already beginning to shake the traditional foundations of the West, especially the Christian religious base. Nietzsche was not unaware of the potential for conflict involved with each of these modes of consciousness. His own emerging or evolving concern, however, was not primarily at least, with species, class, race, or nationality, but rather with sexual or gender consciousness and with historical consciousness, as the latter opened up new possibilities for understanding the former. As Nietzsche's own body and his life were evolving, his *self*-consciousness was evolving, his increasing awareness of his own experiences as an individual, as a male individual, as a male *body*, similar to other male bodies. Also constantly evolving was his sexual or gender consciousness, his awareness of the physiological differences between male and female bodies, and of the possible psychological and social implications of these differences. As I have written before, part of his own genius, and of his distress, was that his consciousness — species, race, class, national, gender, historical — was racing far ahead of those of his contemporaries. It has been noted that Sigmund Freud, who borrowed heavily from Nietzsche, said of him that he probably understood himself better than any man who had ever lived. As to his views regarding experience, this was one expression:

As interpreters of our experiences. — One sort of honesty has been alien to all founders of religions and their kind. They have never made their experiences a matter of conscience for knowledge. "What did I really experience? What happened in me and around me at that time? Was my reason bright enough? Was my will opposed to all deceptions of the senses and bold in resisting the fantastic?" None of them has asked such questions, nor do any of our dear religious people ask them even now. On the contrary, they thirst after things that *go against reason*, and they do not wish to make it too hard for themselves to satisfy it. So they experience "miracles" and "rebirths" and hear the voices of little angels! But we, we others who thirst after reason, are determined to scrutinize our experiences as severely as a scientific experiment — hour after hour, day after day. We ourselves wish to be our experiments and guinea pigs (*GS*, 253).

Walter Kaufmann, commenting on this passage, says that this is quintessential Nietzsche.

We know that Nietzsche was not a typical, conventional, systematic philosopher. Some would not call him a philosopher. His deliberate approach to thinking, his belief that no "truths", no ideas or ideals, no interpretations, no theories, no forms, were sacrosanct, was evidenced clearly in his rejection, or reversal, of philosophical systems in favor especially of aphorisms. These aphorisms — this style or mode of writing — might consist of one, a few, or several sentences, often ending in an ellipsis, intended to be completed by himself or by his reader. He said that his aphorisms were designed to open the reader to further contemplation. So, frequently they remind the reader of one of his reversals, when he wrote, "I don't think thoughts, thoughts think me." This was such an important aspect of his experience. It had the characteristic of demand, often of urgency.

Most of Nietzsche's books did not include an index. Consequently, subjects such as struggle, or conflict, or strife, or war, might show up in any of his works. One must search. The words themselves do appear in his writing. However, in listening to him, in "hearing him with one's eyes", as he suggested, there is confirming evidence throughout that the process of his own life, and that of an inner life devoted to thinking, were processes of struggle. His own beliefs, thoughts, ideas were constantly in competition, in conflict; and ideas were more important than merely items to be bargained for in the "marketplace". They were to be advanced, tested, fought for, defended, resisted. The process was literally a war of ideas and words. And Nietzsche became the exalted warrior.

There are scattered references to the word "war" among Nietzsche's writings. Only as an opening salvo, I have selected a few of his comments. Keep in mind that with Nietzsche every word must be approached with the recognition that he was a master wordsmith who delights seriously in intentional ambiguity. Here, then, are a few of his words, intended to launch us into uncharted, open seas:

> *War.* One can say against war that it makes the victor stupid and the vanquished malicious. In favor of war, one can say that it barbarizes through both these effects and thus makes man more natural; war is the sleep or wintertime of culture: man emerges from it with more strength, both for the good and for the bad (*HA*, 213).

> *Against an enemy.* How good bad music and bad reasons sound when one marches against an enemy! (*PN*, 557)

> *Preparatory human beings.* — I welcome all signs that a more virile, warlike age is about to begin, which will restore honor to courage above all. For this age shall prepare the way for one yet higher, and it shall gather the strength that this higher age will require some day — the age that will carry heroism into the search for knowledge and that will *wage wars* for the sake of ideas and their consequences. To this end we now need many preparatory courageous human beings who cannot very well leap out of nothing, any more than out of the sand and slime of present-day civilization and metropolitanism — human beings who know how to be silent, lonely, resolute, and content and constant in invisible activities; human beings who are bent on seeking in all things for what in them must be *overcome*; human beings distinguished as much by cheerfulness, patience, unpretentiousness, and contempt for all great vanities as by magnanimity in victory and forbearance regarding the small vanities of the vanquished; human beings whose judgment concerning all victors and the share of chance in every victory and fame is sharp and free; human beings with their own festivals, their own working days, and their own periods of mourning, accustomed to command with assurance but instantly ready to obey when that is called for — equally proud, equally serving their own cause in both cases; more endangered human beings, more fruitful human beings, happier beings! For believe me: the secret for harvesting from existence the greatest fruitfulness and the greatest enjoyment is — to *live dangerously!* (*GS*, 228)

> *In media vita.* — No, life has not disappointed me. On the contrary, I find it truer, more desirable and mysterious every year — ever

since the day when the great liberator came to me: the idea that life could be an experiment of the seeker of knowledge — and not a duty, not a calamity, not trickery. — And knowledge itself: let it be something else for others; for example, a bed to rest on, or the way to such a bed, or a diversion, or a form of leisure — for me it is a world of dangers and victories in which heroic feelings, too, find places to dance and play. *"Life as a means to knowledge"* — with this principle in one's heart one can live not only boldly but even gaily, and laugh gaily too. And who knows how to laugh anyway and live well if he does not first know a good deal about war and victory? (*GS*, 255)

. . . I am not afraid of predicting a few things and thus, possibly, of conjuring up the causes of wars (*WP*, 81).

Good luck in fate. — The greatest distinction that fate can bestow on us is to let us fight for a time on the side of our opponents. With that we are *predestined* for a great victory (*GS*, 255).

Out of life's school of war: What does not destroy me, makes me stronger (*PN*, 467).

On War and Warriors

We do not want to be spared by our best enemies, nor by those whom we love thoroughly. So let me tell you the truth!

My brothers in war, I love you thoroughly; I am and I was of your kind. And I am also your best enemy. So let me tell you the truth!

I know of the hatred and envy of your hearts. You are not great enough not to know hatred and envy. Be great enough, then, not to be ashamed of them.

And if you cannot be saints of knowledge, at least be its warriors. They are the companions and forerunners of such sainthood.

I see many soldiers: would that I saw many warriors! "Uniform" one calls what they wear: would that what it conceals were not uniform!

You should have eyes that always seek an enemy — *your* enemy. And some of you hate at first sight. Your enemy you shall seek, your war you shall wage — for your thoughts. And if your thought be vanquished, then your honesty should still find cause for triumph in that. You should love peace as a means to new wars — and the short peace more than the long. To you I do not recommend work but struggle. To you I do not recommend peace but victory. Let your work be a struggle. Let your peace be a victory! One can be silent and sit still only when one has bow and arrow: else one chatters and quarrels. Let your peace be a victory!

You say it is the good cause that hallows even war? I say unto you: it is the good war that hallows any cause. War and courage have accomplished more great things than love of neighbor. Not your pity but your courage has so far saved the unfortunate.

"What is good?" you ask. To be brave is good. Let the little girls say, "To be good is what is at the same time pretty and touching."

They call you heartless: but you have a heart, and I love you for being ashamed to show it. You are ashamed of your flood, while others are ashamed of their ebb.

You are ugly? Well then, my brothers, wrap the sublime around you, the cloak of the ugly. And when your soul becomes great, then it becomes prankish; and in your sublimity there is sarcasm. I know you.

In sarcasm the prankster and the weakling meet. But they misunderstand each other. I know you.

You may have only enemies whom you can hate, not enemies you despise. You must be proud of your enemy: then the successes of your enemy are your successes too.

Recalcitrance — that is the nobility of slaves. Your nobility should be obedience. Your very commanding should be an obeying. To a good warrior "thou shalt" sounds more agreeable than "I will." And everything you like you should first let yourself be commanded to do.

Your love of life shall be love of your highest hope; and your highest hope shall be the highest thought of life. Your highest thought, however, you should receive as a command from me — and it is: man is something that shall be overcome.

Thus live your life of obedience and war. What matters long life? What warrior wants to be spared?

I do not spare you; I love you thoroughly, my brothers in war!

Thus spoke Zarathustra (*Z*, 46, 47, 48).

The more a woman is a woman the more she defends herself tooth and nail against rights in general: for the state of nature, the eternal *war* between the sexes puts her in a superior position by far (*EH*, 76).

Nietzsche was obviously speaking of war with weapons (Hesiod's "evil" Eris) and also of war with words (Hesiod's "good" Eris). For Nietzsche, his weapons were words, carrying his ideas. And he himself was the warrior. He wrote this, "The Germans invented gunpowder — all credit to them! But they made up for it by inventing the printing press." And Nietzsche's friend, Peter Gast, told him, "you have such guns as have never yet existed, and you need only shoot blindly to inspire terror all around." And finally, in what were some of his last words to himself, about himself, and perhaps to future disciples or interpreters, in his book, *Ecce Homo*, Nietzsche wrote:

> I know my fate. One day there will be associated with my name the recollection of something frightful — of a crisis like no other before on earth, of the profoundest collision of conscience, of a decision evoked *against* everything that until then had been believed in, demanded, sanctified. I am not a man. I am dynamite. — ... I was the first to *discover* the truth, in that I was the first to sense — smell the lie as lie... My genius is in my nostrils... For when truth steps into battle with the lie of millennia we shall have convulsions, an earthquake spasm, a transposition of valley and mountain such as has never been dreamed of. The concept politics has then become completely absorbed into a war of spirits, all the power-structures of the old society have been blown into the air — they one and all reposed on the lie: there will be wars such as there have never yet been on earth. Only after me will there be *grand politics* on earth (EH, 126, 127).

Earlier in *The Gay Science* he had written:

> *Who is most influential.* — When a human being resists his whole age and stops it at the gate to demand an accounting, this *must* have influence. Whether that is what he desires is immaterial; that he *can* do it is what matters (GS, 198).

Hesiod's interpretation of the agonistic character of human existence, at the same time, was referring to pairs of opposites — female goddesses and male goddesses, and mortals and immortals — always in conflict or competition. His emphasis was on the struggle.

Heraclitus had similar interests. However, he emphasized the idea of opposites or contraries — Logos, as the controlling principle of the universe. These opposites, however, while never independent of each other, were in a constant state of strife or conflict. Both Hesiod and Heraclitus, as was the case with Nietzsche, saw in war, conflict, strife, the potential for creativity and destruction — as the eternal source of both. So, it appears that in considering Nietzsche and war, we are drawn necessarily to the idea of contraries, to the inextricable relationship between the two.

No single notion, or idea, is more frequently visible in Nietzsche's writings than that of contraries. Recognizing the dangerous nature of his developing criticisms, he wrote this:

> Everything deep loves masks; the deepest things have a veritable hatred of image and likeness. Might not *contrariety* be the only disguise to clothe the modesty of a god? A question worth asking. It would be surprising if some mystic hadn't at some time ventured upon it. There are events of such delicate nature that one would do well to bury them in gruffness and make them unrecognizable. . . Such a concealed one, who instinctively uses speech for silence and withholding, and whose excuses for not communicating are inexhaustible, *wants* and encourages a mask of himself to wander about in the hearts and minds of his friends. And if he doesn't want it, one day his eyes will be opened to the fact that the mask is there anyway, and that it is good so. Every thinker needs a mask; even more, around every deep thinker a mask constantly grows, thanks to the continually wrong, i.e., superficial, interpretations of his every word, his every step, his every sign of life (*BGE*, 46, 47).

Embracing the philosophy of Heraclitus, but typically appealing increasingly to his own experiences, Nietzsche found that the process of his life more than confirmed the reality and significance of opposition. In fact, one might claim that his entire philosophy is a giant mosaic of contraries. For anyone familiar with Nietzsche's writings, here are a few of the most obvious ones: sickness and health, heat and cold, affirmation and negation, ascent and descent, night and day, over and under, same and different, truth and lie, hard and soft, command and obey, benefit and harm, good and evil, strength and weakness, creative and destructive, life and death, truth and error, permanence and change, appearance and reality, life and death, objective and subjective, personal and impersonal, pleasure and pain, beautiful and ugly, good and bad, joy and sorrow, war and peace, male and female.

Important for Nietzsche's experience was the fact that he had frequent episodes of becoming ill, followed by recovery, and then the reversal. He often recounted the decrease and increase in his energy and ability — his powers — that were vividly part of this recurring process. Like night and day, sickness and health were part of life, as were pain and pleasure, joy and sorrow, and much more. One did not have the health and not the sickness, the pleasure and not the pain, the joy and not the sorrow. Having both was a *necessity*; one without the other was never a possibility of nature, of life.

Many opposites make up the structure of the natural world — Heraclitus' Logos. Early in his career Nietzsche wrote:

> When one speaks of *humanity*, the idea is fundamental that this is something which separates and distinguishes man from nature. In reality, however there is no such separation: "natural" qualities and those called truly "human" are inseparably grown together. Man, in his highest and noblest capacities, is wholly nature and embodies its uncanny dual character (*PN*, 32).

Many opposites based within nature and available to sense experience found their way into mythology — perhaps something Nietzsche learned during his long acquaintance with Bachofen. Examples such as night and day, sun and moon, left and right — and the notion that many of these pairs of opposites, used symbolically, referred to male and female. Teasingly perhaps, Nietzsche suggested a few of his own. "Supposing that Truth is a woman — ." Or, "Life is a woman." Or, "Justice is a woman."

As his thinking developed, Nietzsche was obviously and increasingly overwhelmed, amazed, and enchanted by the apparent orderliness of the natural world and its dualistic structure. Opposites, contraries were everywhere. And he adopted, and adapted many of these for his own pleasure and power.

It should be noted that there were a few pairs of opposites which Nietzsche rejected as artificial, inappropriate, invented, misapplied, or false. The pair beginning and end was one. While it was appropriate in many situations, in many events, in a single life, it was not applicable in speaking of the beginning and the end in referring to the universe — neither in philosophical nor theological discourse. Also for Nietzsche, to cast body and soul as opposites was an invention, or false. Since he argued that the soul, either in Platonic thought or Christian thought, was an invention and false, it had no legitimacy in the context of opposites. As we will see later, only the body exists and the soul is "something about the body". As for heaven and hell, they were both fictitious inventions, useful in the power of Christianity, but only extensions of pleasure and pain, as well as reward and punishment.

As for the opposites good and evil, they became the title of one of Nietzsche's most popular books, *Beyond Good and Evil*. The title should suggest that perhaps his interpretation of this pair of contraries was not in agreement with conventional usage and understanding. As we will see later when discussing more thoroughly the subject of values, and as is the case with all values, good and evil are not qualities, or characteristics, that are inherent in individuals or actions. They are subjective and relative to the observer, interpreter, or speaker. Also, good and evil are not abstract entities in themselves. One would need to read his book to understand the many perspectives which he takes in such an expanded and revolutionary interpretation. Humans create or invent these values — good and evil — apply them, and as the title of the book implies, must transcend them in favor of rethinking them. This was part of his anticipated enterprise of the "revaluation of all values".

As a pair of opposites, same and different was one of those which interested Heraclitus — the experience of stepping into the same river, but simultaneously into ever-changing different waters. These contraries, which Nietzsche explores in several ways, become wonderfully fruitful in his somewhat oblique addressing of what I will refer to as the question of "identity" later in the discussion.

As for the opposites hard and soft, Nietzsche used these as symbols — perhaps gleaned from mythological sources, or perhaps from his own imagination. I have interpreted the following short story as a parable, a fable, or perhaps an allegory:

> "Why so hard?" the kitchen coal once said to the diamond. "After all, are we not close kin?" Why so soft? O my brothers, thus I ask you: are you not after all my brothers?
>
> Why so soft, so pliant and yielding? Why is there so much denial, self-denial, in your hearts? So little destiny in your eyes?
>
> And if you do not want to be destinies and inexorable ones, how can you one day triumph with me?
>
> And if your hardness does not wish to flash and cut and cut through, how can you one day create with me?
>
> For creators are hard. And it must seem blessedness to you to impress your hand on millennia as on wax.

Blessedness to write on the will of millennia as on bronze —
harder than bronze, nobler than bronze. Only the noblest is altogether hard.

This new tablet, O my brothers, I place over you: *become hard!* (Z,
214)

These symbols hard and soft offer the persuasive possibility that
Nietzsche was referring to male and female, to future creators of a new
"table of values", the "revaluation of all values", which became increasingly important for him and for us. I would add this notation from F.M.
Cornford, in his book, *From Religion to Philosophy.* In probing the origins
of Greek philosophy, he said, "If we look more closely into this conception of pairs of contraries, . . . we can discern that the prototype of all
opposition or contrariety is the contrariety of sex."

Two major pieces of the puzzle of war — *conflict* or *competition,* and
contrariety — appear to be drawn together like magnets. It seems, then,
that the appropriate next idea to require careful scrutiny is that with
which many people identify Nietzsche — *power.*

Life, as Nietzsche interpreted it and lived it, was throughout necessarily *agonistic* and *dualistic.* His dualism, as might be expected, was
not that of traditional views, not a theory stressing two irreducible
elements, e.g., matter and spirit, body and soul, or two opposing principles, one of which is good and the other evil. Evidence and experience
left no doubt that the underlying, overriding dualistic nature of the
universe consisted of the prototype of contrariety — sexuality, female
and male, two forms, two bodies, separate but interdependent. It became increasingly apparent to Nietzsche that the prototype of the agon
— competition and conflict — corresponded with that of contrariety.
Competition and conflict between the sexes. Much more of this later.

To the ideas of life understood as agonistic, and as dualistic,
Nietzsche contributed his famous "life as will-to-power." Earlier it was
suggested that Nietzsche believed, Hobbes believed, and before Hobbes, Machiavelli believed, that the basic starting point, the center of all
philosophy and science was *power.* All other ideas seemed to be generated from, or refer back to, this base. As a political philosopher, Machiavelli's concerns were to investigate and elaborate the appropriate and
effective means of acquiring, maintaining, and increasing the power, or

powers, of the ruler, and that meant the state. The threat of force, or the use of force, were the bases of the ruler's power. For one who rules, the end remains the same — power — justifying any means. Three particular instruments, or means, receive special attention by Machiavelli — fear, cruelty, and deceit. However, in acquiring, maintaining, and expanding his power, the major means of the ruler is *war*. Recall this:

> A prince should therefore have no other aim or thought, nor take up any other thing for his study, but war and its organization and discipline, for that is the only art that is necessary to one who commands, and it is of such virtue that it not only maintains those who are born princes, but often enables men of private fortune to attain to that rank (*GPT*, 291).

As a reminder, Hobbes saw man as naturally and essentially combative and competitive — agonistic. And, the compelling concern of that conflict or competition, which he saw as the identifying characteristic of *every man*, is the desire for power, or powers. Concerned with bodies in motion, Hobbes' focus was the human body, and the continuous motion of most interest was the motion of conflicting, competing, combatting.

Whereas Machiavelli appears to have developed his interpretation solely on the experiences of the ruler, Hobbes expanded his interpretation to include every living person. And he also gave special attention, influenced by Machiavelli, to the ruler, who in the pursuit of power "makes laws at home and wars abroad." Hobbes emphasized that men compete for riches, for honor, for command, and in the process are inclined toward "contention, enmity, and war" — and to the attainment of the desire to "kill, subdue, supplant, or repel the other."

Nietzsche gathered much from both Machiavelli and Hobbes and their approaches to power. As with Hobbes, Nietzsche's interpretation included all living persons. The most explosive and amazing idea from Nietzsche, however, was that living persons come in two separate forms. History and Bachofen's mythology, had reassured him that this had always been the case — male bodies and female bodies, competing in the pursuit of power.

Just as conflict or competition, and contrariety are inextricably related, so *power* and *war* are similarly connected. The topic now is Nietzsche's interpretation of power, with him speaking philosophically, psychologically, and in his terms, "politically." Later I will suggest my interpretation of Nietzsche's interpretation.

But before hearing from Nietzsche, a closer look at the idea and term "power" seems to be in order. I suggest that the phenomenon of power is probably the most complex, complicated, intricate, baffling, exciting, dangerous, and important element in human experience. The Encyclopaedia Britannica offers a separate and limited account of energy and of power — both of which were important for Nietzsche. First these:

> [E]nergy, in physics, the capacity for doing work. It may exist in potential, kinetic, thermal, electrical, chemical, nuclear, or other various forms. There are, moreover, heat and work — i.e., energy in the process of transfer from one body to another. After it has been transferred, energy is always designated according to its nature. Hence, heat transferred may become thermal energy, while work done may manifest itself in the form of mechanical energy.

> All forms of energy are associated with motion. For example, any given body has kinetic energy if it is in motion. A tensioned device such as a bow or spring, though at rest, has the potential for creating motion; it contains potential energy because of its configuration. Similarly, nuclear energy is potential energy because it results from the configuration of subatomic particles in the nucleus of an atom.

> Energy may be converted from one form to another in various ways. Usable mechanical or electrical energy is, for instance, produced by many kinds of devices, including fuel-burning heat engines, generators, batteries, fuel cells, and magnetohydrodynamic systems.

> [E]nergy, conservation of, principle of physics according to which the energy of interacting bodies or particles in a closed system remains constant. The first kind of energy to be recognized was kinetic energy, or energy of motion. . . . The notion of energy was progressively widened to include other forms.

> [P]ower, in science and engineering, time rate of doing work, or delivering energy, expressible as the amount of work done W, or energy transferred, divided by the time interval t — or W/t. . . . Power is expressible also as the product of the force applied to move an object and the speed of the object in the direction of the force.

> [P]ower, balance of, in international relations, the posture and policy of a nation or group of nations protecting itself against another nation or group of nations by matching its power against

the power of the other side. States can pursue a policy of balance of power in two ways: by increasing their own power, as when engaging in an armaments race or in the competitive acquisition of territory; or by adding to their own power that of other states, as when embarking upon a policy of alliances.

Webster's Encyclopedic Unabridged Dictionary of the English Language (Copyright 1996 by Random House Value Publishing, Inc.) adds noticeably in expanding the possible meanings of energy and power in the direction which Nietzsche would take. The following are important:

[E]nergy — capacity for vigorous activity; available power; adequate or abundant amount of such power; a feeling of tension caused or seeming to be caused by such power; an exertion of such power; the ability to act, lead others, affect forcefully; forcefulness of expression; physics, capacity to do work; any source of usable force, as fossil fuel, electricity, or solar radiation

[P]ower — ability to do or act; capability of doing or accomplishing something; political or natural strength; great ability to do or act; possession of control or command over others; authority; ascendancy; political ascendancy or control in government, state, country; person or thing that possesses or exercises authority or influence; power or powers (physics, math, optics); state or nation having international authority or influence; military or naval force; physics, work done or energy transferred per unit of time; mechanical energy; energy, force, or momentum; to inspire, spur, sustain; expressing or exerting power

[S]trength — quality or state of being strong; bodily or muscular power; mental power, force, vigor; moral power; firmness or courage; power by reason of influence, firmness, or courage; effective force, potency, or cogency, as of inducements or arguments; power of resisting force, strain, wear; vigor of action, language, feeling, etc.; inherent capacity to manifest energy, to endure, and to resist might — power or strength in great or overwhelming degree (e.g., of army).

These, then, are some of the many meanings which attach to the idea, or word, *power*: energy, ability, capability, capacity, strength, influence, might, force, inspire, persuade, arouse, authority, domination, control, command. *Capability* suggests the potential or ability to act, produce, generate, create (and destroy). *Authority* suggests the power to rule, to direct the actions or thoughts of others, usually because of *rank*, to issue commands and punish violations. *Control* suggests power or influence applied to complete or successfully complete, regarding

persons or actions — usually implying the requirement of fear, weakness or obedience. *Influence* suggests personal and unofficial power derived from deference of others to one's character, ability or rank — the
power to arouse, inspire, excite, incite, persuade, impress, directly affect opinions, ideas, taste. Nietzsche had the *power* — the energy, the
ability, to destroy and create ideas and *words*, to *influence* others.

Nietzsche's philosophical and philological interests and inqui-
ries took him back to the early philosophers of nature, the so-called
"Pre-Socratics", to whom he was especially attracted. From the limited
fragments of these philosophers, it is evident that many of their ideas
or interpretations became foundational for Nietzsche's philosophy of
power and for the further development of much of his thinking. What
had been emphasized by some of these earliest philosophers was their
agreement, with varying ways of expressing their views, that motion —
the eternal process of change — was the dominant feature of our natural world. Material bodies in constant motion, from the most minute in
size to bodies of many sizes and types — that was the starting point.
For Nietzsche, not surprisingly, the focus became the constant motion,
or change, having to do with human bodies. His deeper thinking and
interpretations required an attempt to answer the question — what is
the source, or origin, of motion, all motion? Of this process of change?

Recall, Heraclitus had pointed out one kind of polar opposition —
that which consists in the regular succession by one thing of its opposite, so that if one perished, so would the other. He noted, "The cold
things get hot, hot gets cold, wet gets dry, dry things get damp." Or,
the alternation between night and day, or sun and moon, and numerous others. The key idea for Nietzsche became, not that there are these
opposites — day and night, hot and cold, wet and dry, and others, but
that each one of any pair is in the continuing process of *becoming* the
other. The question was how are we to understand this eternally recurring process. In the natural world this is motion, this is change. Everything in the universe is involved in the constant process of becoming.
The question, then, was what can explain this process, and Nietzsche's
answer was as follows:

> And do you know what "the world" is to me? Shall I show it to
> you in my mirror? This world: a monster of energy, without be-
> ginning, without end; an immovable, brazen enormity of energy,

which does not grow bigger or smaller, which does not expend itself but only transforms itself; as a whole of unalterable size, a household without expenses or losses, but likewise without increase or income; enclosed by "nothingness" as by a boundary; not something flowing away or squandering itself, not something endlessly extended, but as a definite quantity of energy set in a definite space and not a space that might be "empty" here or there, but rather as energy throughout, as a play of energies and waves of energy at the same time one and many, increasing here and at the same time decreasing there; a sea of energies flowing and rushing together, eternally moving, eternally flooding back, with tremendous years of recurrence, with an ebb and flow of its forms; out of the simplest forms striving towards the most complex, out of the stillest, most rigid, coldest form towards the hottest, most turbulent, most self-contradictory, and then out of this abundance returning home to the simple, out of the play of contradiction back to the joy of unison, still affirming itself in the uniformity of its courses and its years, blessing itself as that which must return eternally, as a becoming that knows no repletion, no satiety, no weariness — : this is my *Dionysian* world of the eternally self-creative, the eternally self-destructive, this mystery world of the twofold delight, this my "beyond good and evil", without aim, unless the joy of the circle is itself an aim; without will, unless a ring feeling goodwill towards itself — do you want a *name* for this world? A *solution* for all your riddles? A *light* for you too, you best concealed, strongest, least dismayed, most midnight men? — *This world is the will to power — and nothing beside!* And you yourself are also this will to power — and nothing beside! (*N*, 136)

In this memorable interpretation, the controlling, defining principle of the universe is energy, power. One thing, energy is ubiquitous — existing everywhere at the same time. All movements, all activities, require energy. And Nietzsche gives a name to this energy — "the will to power". The world is will to power. As he says, this is his definition. Life, to be alive, is to be the will to power. Man is the will to power. And this was his focus — the human being, constantly changing, in motion, becoming. Not atoms in motion, as Democritus had suggested, but bodies in motion — and most importantly, human bodies. And human bodies occur in two forms, and present themselves in multiple, indeterminate motions, or activities.

Nietzsche's "will to power" is a natural impulse, a drive, a feeling, a need to discharge, express, exercise, increase this energy. It is ambiguous. Humans constantly experience this process, these motions, change, power in different forms, different ever-shifting patterns. All human activity is infused with this drive for power, for the "feeling of power". Nietzsche urged us to experiment with the idea that the will to power could explain *all* human activities. And those activities which he obviously considered fundamental were those of the five senses — seeing, hearing, tasting, smelling, and touching. The functions of the eyes, the ears, the mouth, the nose, and the hands. All of these are active in having access to the world in which humans are participants, in the universe of change, process, becoming. They are the physiological aspects of human bodies, the powers, abilities, which enable humans in constantly engaging this enormous, ubiquitous, constantly changing world — natural and social or cultural. Nietzsche's passion, his intellect, and his wit pursued with vigor what is often referred to merely as "sense experience". He thought and wrote about all of the senses, but perceiving had by far most of his attention. And as with all of his ideas, the point of reference was always power.

By taking the idea of power "globally", so to speak, Nietzsche presented himself almost literally with endless opportunities, or perspectives, to suggest what that might mean. His attention remained focused within the natural world, the world of *nature*, as he perceived and reflected on it. For example, the energy exhibited by the four elements — air power, fire power, water power, and earth power. Or, the power of the five senses.

For Nietzsche the vital force, the impulse to life, the "élan vital", that which he described as the basic nature of all living things, was this impulse or drive for power, what he called "will to power". He believed it would be possible, and attempted to understand all human behavior by experimentally applying this idea to countless examples of human activity. His own rich experiences, personal and historical, were his evidence. Having power, the feeling of power, preserving and enhancing power, became in his thinking the sole motive for humans. The differences in behavior were a matter of means to the end — power. Only the means might be judged, not the end. A "common basic drive" unites the most diverse activities. This is the fundamental fact of life, for every living creature.

Human beings, persons and the extraordinary possible powers associated with human bodies — that was Nietzsche's focus. *Life*, the actual biological, physiological process of producing new life, more life; sexuality, the basic fact, the first premise, the presupposition of life — female and male bodies — the "uncanny duality of nature". The basic expression, or exercise, of the will to power is creativity, the power of generation — taken in its broadest sense. Beyond creating life, there is little doubt that Nietzsche most valued the thoughts, and the ability to communicate those thoughts in words, the power of speaking and writing — the *power of words*. In fact, to create and speak words about the physiological process of creating life — of all creating and generating.

One of Nietzsche's best translators and interpreters, R. J. Hollingdale, wrote this:

> To grasp Nietzsche's theory of will to power and its ramifications one cannot do better than trace the idea as it appears at this or that place in his works and see how it formed into a hypothesis which was then consciously employed, consistently yet still experimentally, as an explanatory principle (N, 76).

In his Preface to his book entitled *Nietzsche*, Hollingdale says, "Interpretation is ultimately no more than a necessary method of presentation: the book's objective is to make the reader want to read Nietzsche for himself. If it succeeds in that objective, the reader will quickly become his own 'interpreter'." I know of no more persuasive interpretation of Nietzsche's philosophy of power than that of Hollingdale.

In the early stages of developing his doctrine of the will to power, Nietzsche wrote this:

> *On the doctrine of the feeling of power:* — Benefiting and hurting others are ways of exercising one's power over others; that is all one desires in such cases. One hurts those whom one wants to feel one's power, for pain is a much more efficient means to that end than pleasure; pain always raises the question about its origin while pleasure is inclined to stop with itself without looking back . . . Certainly the state in which we hurt others is rarely as agreeable, in an unadulterated way, as that in which we benefit others; it is a sign that we are still lacking power, or it shows a sense of frustration in the face of this poverty; it is accompanied by new dangers and uncertainties for what power we do possess, and clouds our horizon with the prospect of revenge, scorn, punishment and failure (GS, 86, 87).

Nietzsche is suggesting here that hurting others is a sign that one lacks power — something of a contrary position from conventional belief. Creating or destroying are ways of exercising one's power, enhancing or diminishing life (that of oneself or of others), affirming or denying one's own fate, and especially as we will see later, destroying old values and creating new ones.

There is little doubt that Nietzsche, in his interpretation, reached the possibility that there exists a different exercise of power. Power over oneself, command of oneself, rather than over others. And the possibility of creating oneself, considering as he says, that we are both "creature" and "creator". Read a typical passage from *The Gay Science*:

> *One thing is needful.* — To "give style" to one's character — a great and rare art! It is practiced by those who survey all the strengths and weaknesses of their nature and then fit them into an artistic plan until every one of them appears as art and reason and even weaknesses delight the eye. Here a large mass of second nature has been added; there a piece of original nature has been removed — both times through long practice and daily work at it. Here the ugly that could not be removed is concealed; there it has been reinterpreted and made sublime. Much that is vague and resisted shaping has been saved and exploited for distant views; it is meant to beckon toward the far and immeasurable. In the end, when the work is finished, it becomes evident how the constraint of a single taste governed and formed everything large and small. Whether this taste was good or bad is less important than one might suppose, if only it was a single taste!. . . For one thing is needful: that a human being should attain satisfaction with himself, whether it be by means of this or that poetry and art; only then is a human being at all tolerable to behold. Whoever is dissatisfied with himself is continually ready for revenge, and we others will be his victims, if only by having to endure his ugly sight. For the sight of what is ugly makes one bad and gloomy (*GS*, 232, 233).

Near the beginning of one of Nietzsche's last works, The *Antichrist*, we find these questions and answers:

> What is good? — All that heightens the feeling of power, the will to power, power itself in man. What is bad? — All that proceeds from weakness. What is happiness? — The feeling that power *increases* — that a resistance is overcome (N, 76).

And speaking of these questions and answers, Hollingdale says:

> The abruptness and uncompromisingness of these assertions are characteristic of the works of 1888: they are a condensation of the

sense and purpose of the whole "philosophy of power" as it has been built up during the course of years (N, 76).

In Part Two of this book I provided a fairly extensive account of Thomas Hobbes' lengthy analysis of power. As C.B. Macpherson claimed regarding Hobbes, "He exposed the lineaments of power more clearly than anyone had done since Machiavelli, more systematically than anyone had ever done, and than most have done since." Nietzsche, as we know, took the demands to understand power and power relations to new heights, to new and dangerous depths.

But in addition to having Hobbes' brilliant analysis of power and powers upon which to draw, there was something else which, I would argue, became for Nietzsche equally, perhaps ultimately more, important. Hobbes was one of the first philosophers to address and question directly the subject of value, or values. After specifying and analyzing the primary powers of a man and the central place of power in the fabric of life, Hobbes states firmly:

> *Worth.* The *value*, or WORTH of a man, is as of all other things, his price; that is to say, so much as would be given for the use of his power: and therefore is not absolute; but a thing dependant on the need and judgment of another. An able conductor of soldiers, is of great price in time of war present, or imminent; but in peace not so. A learned and uncorrupt judge, is much worth in time of peace; but not so much in war. And as in other things, so in men, not the seller, but the buyer determines the price. For let a man, as most men do, rate themselves at the highest value they can; yet their true value is no more than it is esteemed by others (*L*, 73).

There are two additional perspectives taken by Hobbes regarding values, both of which were enormously important as Nietzsche shaped his own interpretations. First, recall that Hobbes wrote this:

> *Good. Evil.* But whatsoever is the object of any man's appetite or desire, that is it which he for his part calleth *good*: and the object of his hate and aversion *evil*; and of his contempt, *vile* and *inconsiderable*. For these words of good, evil, and contemptible, are ever used with relation to the person that useth them: there being nothing simply and absolutely so; nor any common rule of good and evil, to be taken from the nature of the objects themselves; but from the person of the man, where there is no commonwealth; or, in a commonwealth, from the person that representeth it; or from an arbitrator or judge, whom men disagreeing shall by consent set up, and make his sentence the rule thereof (*L*, 48, 49).

This notion that the values "good" and "evil", and by implication other values, are not intrinsic to the nature of objects themselves, but rather "belong to", or are assigned by the person on the basis of desire or aversion, would become an absolutely central idea for Nietzsche. And as we will see, explosive and threatening to Western society and culture — in the hands of Nietzsche.

Second, in exploring and developing what I think was new ground for Hobbes, he began, as we have shown, with the idea that the value of a person was determined, not by the person himself, but by some other, or others. And Hobbes' view was that this value was based on a quantitative calculation — the *price*. It was, as with everything else, a question of the marketplace — buying and selling of one's power, or powers. Also, the price would vary with the times — as he suggested, times of war and times of peace.

The power, or powers, of a man which attracted and were considered desirable by the "buyer", in Hobbes' analysis, were these: extraordinary strength, form, prudence, arts, eloquence, liberality, nobility, riches, reputation, friends, and good luck. However, he recognized that the market calculus was less than perfect, and continued with this:

> *Worthiness. Fitness.* WORTHINESS, is a thing different from the worth, or value of a man; and also from his merit, or desert, and consisteth in a particular power, or ability for that, whereof he is said to be worthy: which particular ability, is usually named FITNESS, or *aptitude*.

> For he is worthiest to be a commander, to be a judge, or to have any other charge, that is best fitted, with the qualities required to the well discharging of it; and worthiest of riches, that has the qualities most requisite for the well using of them: any of which qualities being absent, one may nevertheless be a worthy man, and valuable for something else. Again, a man may be worthy of riches, office, and employment, that nevertheless, can plead no right to have it before another; and therefore cannot be said to merit or deserve it (*L*, 79).

The philosophical contributions of Hobbes to the subject of power are well known and have been the source of a wealth of critical thinking and writing. The subject of *values*, of his handling of the subject, may have been equally momentous. Certainly one may make the case that this was so with Nietzsche. When Nietzsche takes the question of the value of a person, or persons, and the relation of such value, or lack of it, to power or powers, this becomes the basis of his famous challenge for mankind — the *"revaluation of all values"*. This undoubtedly will provide the key to his possible thinking regarding *war*.

In its biography of Nietzsche, the Encyclopaedia Britannica says this:

> In his mature writings Nietzsche was preoccupied by the origin and function of values in human life. If, as he believed, life neither possesses nor lacks intrinsic value and yet is always being evaluated, then such evaluations can usefully be read as symptoms of the condition of the evaluator. He was especially interested, therefore, in a probing analysis and evaluation of the fundamental cultural values of Western philosophy, religion, and morality, which he characterized as expressions of the ascetic ideal.

To grasp the significance of values in Nietzsche's thinking, one cannot do better than to try to show, although irregularly and imperfectly, how this significance developed to finally become the ultimate driving issue. To begin, I have selected a few quotations from both his earlier writings and his later ones, hoping that by so doing we may begin to have a clearer sense of just what was the presence and force of this idea almost from the beginning.

> When may we begin to *"naturalize"* humanity in terms of a pure, newly discovered, newly redeemed nature? (*GS*, 169)

> Whenever we encounter a morality, we also encounter valuations and an order of rank of human impulses and actions. These valuations and orders of rank are always expressions of the needs of a community and herd; whatever benefits it most — and second most, and third most — that is also considered the first standard for the value of all individuals (*GS*, 174).

> We who think and feel at the same time are those who really continually *fashion* something that had not been there before: the whole eternally growing world of valuations, colors, accents, perspectives, scales, affirmations, and negations.. . . Whatever has value in our world does not have value in itself, according to its nature — nature is always value-less, but has been *given* value at some time, as a present — and it was *we* who gave and bestowed it. Only we have created the world *that concerns man!* — But precisely this knowledge we lack, and when we occasionally catch it for a fleeting moment we always forget it again immediately; we fail to recognize our best power and underestimate ourselves, the contemplatives, just a little. We are *neither as proud nor as happy* as we might be (*GS*, 241, 242).

I see nobody who ventured a *critique* of moral valuations; I miss even the slightest attempts of scientific curiosity, of the refined, experimental imagination of psychologists and historians that readily anticipates a problem and catches it in flight without quite knowing what it has caught. I have scarcely detected a few meager preliminary efforts to explore the *history of the origins* of these feelings and valuations (which is something quite different from a critique and again different from a history of ethical systems).. . . Thus nobody up to now has examined the *value* of that most famous of all medicines which is called morality; and the first step would be — for once to *question* it. Well then, precisely this is our task — (*GS*, 284, 285).

We are far from claiming that the world is worth *less;* indeed it would seem laughable to us today if man were to insist on inventing values that were supposed to *excel* the value of the actual world. This is precisely what we have turned our backs on as an extravagant aberration of human vanity and unreason that for a long time was not recognized as such (*GS*, 286).

On the origin of religions. — The distinctive invention of the founders of religions is, first: to posit a particular kind of life and everyday customs that have the effect of a *disciplina voluntatis* and at the same time abolish boredom — and then: to bestow on this life style an *interpretation* that makes it appear to be illuminated by the highest value so that this life style becomes something for which one fights and under certain circumstances sacrifices one's life. Actually, the second of these two inventions is more essential. The first, the way of life, was usually there before, but alongside other ways of life and without any sense of its special value. The significance and originality of the founder of a religion usually consists of his *seeing* it, *selecting* it, and *guessing* for the first time to what use it can be put, how it can be interpreted (*GS*, 296).

A *revaluation of all values*, this question mark, so black, so tremendous that it casts shadows upon the man who puts it down — such a destiny of a task compels one to run into the sun every moment to shake off a heavy, all-too-heavy seriousness. Every means is proper for this; every "case" a case of luck. Especially *war*. War has always been the great wisdom of all spirits who have become too inward, too profound; even in a wound there is the power to heal. . . . Even the most courageous among us only rarely has the courage for that which he really knows (*PN*, 465, 466).

The problem I thus pose is not what shall succeed mankind in the sequence of living beings (man is an *end*), but what type of man shall be *bred*, shall be *willed*, for being higher in value, worthier of life, more certain of a future (*PN*, 570).

There is no disputing that early in his philosophical development Nietzsche began to give his attention to value, or values. Challenging most of his philosophical contemporaries, he voiced his disagreement with their dominant interest in theories of knowledge, to proclaim instead that not knowledge, but rather values, should be given the highest priority with philosophers. That position, in fact, ultimately clearly defined his own task. From very early, Nietzsche was struck with what he perceived as the condition of German culture, and beyond that, of Western culture. His description — Western culture was deteriorating, descending, decomposing, decadent. Typical of his sometimes sardonic expressions, he wrote:

> We Europeans confront a world of tremendous ruins. A few things are still towering, much looks decayed and uncanny, while most things already lie on the ground. It is all very picturesque — where has one ever seen such beautiful ruins? — and overgrown by large and small weeds. The church is this city of destruction. We see the religious community of Christianity shaken to its lowest foundations; the faith in God has collapsed; the faith in the Christian-ascetic ideal is still fighting its final battle. An edifice like Christianity that had been built so carefully over such a long period — it was the last construction of the Romans! — naturally could not be destroyed all at once. All kinds of earthquakes had to shake it, all kinds of spirits that bore, dig, gnaw and moisten have had to help (*GS*, 310).

Nietzsche's interpretation of the cause, or source, of this condition apparently seemed to him to be self-evident — the *values, decadent values*. More about this immediate situation and concern later. What was demanded was a thorough, deep, intensive, even perhaps dangerous, investigation of values, together with the courage and honesty to conduct such an investigation. Nietzsche was a philosopher of nature, a philosopher of power, but his greatest legacy was his critique of values — to later thinkers and to the world. I have no doubt that questions surrounding values would have been the basis of any critique of war that he might have developed. Certain of his perspectives and interpretations are made very clear. First, the question of the *origin of values*:

> We who think and feel at the same time are those who really continually *fashion* something that had not been there before: the whole eternally growing world of valuations, colors, accents, perspectives, scales, affirmations, and negations.. . . Whatever has *value* in our world now does not have value in itself, accord-

ing to its nature — nature is always value-less, but has been *given* value at some time, as a present — and it was *we* who gave and bestowed it. Only we have created the world *that concerns man!* — But precisely this knowledge we lack, and when we occasionally catch it for a fleeting moment we always forget it again immediately; we fail to recognize our best power and underestimate ourselves, the contemplatives, just a little. We are *neither as proud nor as happy* as we might be. (*GS*, 241, 242)

With this interpretation — the denial of the objective and absolute status of all values, and the affirmation of the subjective and relative status of all values — Nietzsche establishes the rock upon which he builds his extensive formidable critique of values.

After replying to the question of the origin of values, there follows the question of the function, or significance, of values. And Nietzsche's unequivocal answer is in the above quotation — "Only we have created the world *that concerns man!*" Values, of every possible kind, *is* the structuring principle of the world of human beings. Values — which are constantly being created, recreated, changing. The world of human beings, natural or cultural, is a continuing process of valuing, devaluing, and revaluing — by human beings. The pronoun "we" in Nietzsche's statement is a general assignation. But at the same time it is specific. In Nietzsche's perspective, the "value-creators" — the "contemplatives", as he calls them — we understand later, as being theologians and/or philosophers. And in Nietzsche's experience these two categories frequently overlap, are absorbed one into the other, one becomes the other. And notice that Nietzsche recognizes this creative activity as "our best power." In the West, philosophers such as Plato and Aristotle, and theologians such as St. Augustine and St. Thomas Aquinas, have exercised their power in shaping Western culture. Nietzsche did believe that the task of future philosophers was to create, recreate, revalue all values — those values which had brought his world, Western culture, to near death. Obviously, this exercise or kind of power was of the highest value. As we know, Nietzsche considered Western culture as having been the creation/invention of Christian theologians, and they and their accomplices had been successful in their creating. Nietzsche's devastating and notorious attack on Christianity was centered on its moral values. My earlier book, *Nietzsche, Philosopher of the Perilous Perhaps*, explored this in detail, and we will have more to say on the subject later.

Any place, at any time one looks there are values. They litter our world. Value, as a verb, may be understood as meaning to consider with regard to excellence, usefulness, importance, or to regard or esteem highly. Value, as a noun, may be understood as meaning that quality of anything (or any person) which renders it desirable, useful, or important. The term "worth" usually implies spiritual qualities of mind and character, or the moral excellence — of a person. To make a "value judgment", ordinarily considered subjective, is regarded as estimating the worth, quality, or goodness of something or someone.

And, these ubiquitous entities called "values" are indeterminately typed. Aesthetic, ethical, moral, political, religious, economic, scientific, social, mathematical, logical, military, linguistic, educational, environmental, legal, health, property, nutritional, recreational, and more. Nietzsche spoke often of the notion that every culture has its own "table of values" — but never fixed or permanent. For Nietzsche, his philosophical position in which values had top priority was as unshakeable as his position regarding the significant role of values in any culture. Moral values — *morality* — were the basis of all of the others. A careful reading of Nietzsche's writings supports this position. He gives his definition of morality, elaborates on that definition, and confounds most readers when he refers to himself many times as "the first immoralist."

Look first at what is his brief precise definition. In *Beyond Good and Evil*, he wrote:

> In all willing, then, there is commanding and obeying on the basis, as we have seen, of a social structure of many "souls." This is why a philosopher should consider himself justified in including willing within the general sphere of morality — morality understood as the doctrine of the rank-relations that produce the phenomenon we call "life." — (*BGE*, 21, 22)

I have insisted before that this definition — taking into account many other things which preceded and followed it in his writings — surely refers to the rank-relations between females and males. Also, of course, the power-relations and the value-relations. *Value, power,* and *rank* are interlocking fundamentals of our world, "the world *that concerns man!*"

There is no systematic treatment of the subject of values, no "theory of values", in Nietzsche's thinking to which we may appeal, with which we may agree or disagree. However, I ardently believe that any critique of war which Nietzsche might have developed would have reflected his ultimate concern — values, in particular moral values. He makes perfectly clear that his devastating critique of Christianity was centered around its "decadent morality." And there will be much more that needs to be said later in that regard. But, my continuing probe into the subject of values in Nietzsche's thinking will be uncharted and erratic. What does seem like a reasonable assumption is that his, and our, concerns with values would radiate in many directions toward his interests in contraries, religion, physiology, psychology, nature, culture, sexuality, mythology, and history.

Touch base one more time with Nietzsche, and then back into history. One of his last books, *Beyond Good and Evil*, was about morality, and we will return to it often. But right now this is important:

> It is obvious that the moral value-characteristics are at first applied to *people* and only later, in a transferred sense, to *acts* (*BGE*, 203).

In my earlier book, *Nietzsche, Philosopher of the Perilous Perhaps*, in addressing Nietzsche's treatment of the significance of sexuality in Christianity, I wrote this:

> Nietzsche believed that one had only to *choose* to open one's eyes to observe that the producers of "life" — natural or cultural — were males and females. All other distinctions — race, ethnicity, class, etc. — were derivative and secondary. Also, it was clearly perceivable that the rank-relations between the sexes had been established from the very beginning of Christian ideology. Also, there was no disputing that it was clear that this relationship was such as to designate the male as superior and dominant, the female as inferior and submissive. In Nietzsche's view, this arrangement was not an incidental element of Christianity. It was, and continued to be, the core, the essential and enduring element, the aim of this religion from its inception. It was the power which had been intended, sought, and achieved, incredibly successfully.

Nietzsche was aware, however, that this way of thinking probably was not entirely original. Recall this from *Beyond Good and Evil*:

> What is astonishing about the religiosity of the ancient Greeks
> is the lavish abundance of gratitude that radiates from it. Only a
> very distinguished type of human being stands in *that* relation to
> nature and to life. Later, when the rabble came to rule in Greece,
> *fear* choked out religion and prepared the way for Christianity
> (*BGE*, 58).

Perhaps the "rabble" to whom he was referring were Socrates,
Plato, and Aristotle. Nietzsche had shown his distaste, even scorn, for
both Socrates and Plato. But what about Aristotle? His role in preparing the way for Christianity was major. And not only in preparing the
way, as we shall see. Nietzsche could hardly have been unfamiliar with
Aristotle's writings — one of those "dogmatic" philosophers.

Aristotle was the first major Western theoretician to address directly, and at length, the relative status of the female and the male. Considering the overall magnitude of Aristotle's influence on the development
of Western ideas and culture, including especially religious thinking, it
seems apparent that at least a brief incursion into this particular analysis should be ventured. Of particular interest for this issue are first, his
biological writings, especially "The Generation of Animals," in which
he analyzes the respective roles of the male and the female in the reproduction of life. And second his political writings, *Politics*, in which his
concern is the social and political status of the woman. The first major
figure to offer a new "scientific biology", extraordinarily detailed, if ex-
travagantly contrived and convoluted, Aristotle maintains that woman
is a mutilated and incomplete man — a thesis that has had a long and
persistent history. His theory depends on his claim that women have
less "soul" than men. What this comes to is that the male contributes
the "form", or soul, while the female contributes merely the "matter", or
nutrition. That is, the male creates human life, not the female. Having
established, in his view, the authoritative position regarding the obviously deficient natural state of the woman, Aristotle proceeds to argue
that she is physically weaker and is less capable of rational thought.
All to reassure that the subordinate position of the female is justifiably
necessary and rational.

It might be relatively persuasive to claim that more than any other
figure, Aristotle was the "father" of Western intellectual and cultural
history — that is, considering the last two and a half millennia. He an-

nounced the emergence of the patriarchal mode of analysis, interpretation, and speaking. And he did this brilliantly. What followed were,
in a sense, footnotes, annotations, elaborations. His thinking was ac-
claimed, celebrated, embraced, rejected, denied, rebutted, but rarely
ignored. A notable ranking member of Aristotle's genealogy was the
thirteenth-century theologian, Thomas Aquinas. Aristotle was the dar-
ling of Aquinas. In the earlier book I wrote:

> By the thirteenth century the thinking about woman had been
> well established in the medieval church by Paul and Augustine.
> However, the real heavyweight in Christian theology was Aqui-
> nas, with his remarkable synthesis between the Judeo–Christian
> and Greek traditions, especially Aristotle. Aquinas agreed with
> Aristotle's perspective that woman is a defective male, lacking
> in vital force, a "misbegotten", "deformed", "contemptible", crea-
> ture. From a Christian perspective, God created woman in the
> divine order, and her deformity is therefore natural. Aristotle's
> "scientific biology" in no way contradicted Aquinas' "biblical bi-
> ology". The details of Aquinas' interpretation of the validity and
> morality of woman's natural inferiority and social subordination
> are as contrived and convoluted as those of his esteemed and be-
> loved philosopher. The *Summa Theologica* of Aquinas was destined
> to provide the Roman Catholic Church with its official theologi-
> cal and philosophical dogma for centuries to come. And that in-
> cluded its treatment of women, one major aspect of which was
> the silencing of the female voice.

In the early pages of *Beyond Good and Evil*, before he laid down in
fifteen words his definition of morality, here is a lengthy passage which
needs repeating:

> Gradually I have come to realize what every great philosophy
> up to now has been: the personal confession of its originator, a
> type of voluntary and unaware memoirs; also that the moral (or
> amoral) intentions of each philosophy constitute the protoplasm
> from which each entire plant has grown. Indeed, one will do well
> (and wisely), if one wishes to explain to himself how on earth
> the more remote metaphysical assertions of a philosopher ever
> arose, to ask each time: What sort of morality is this (is *he*) aim-
> ing at? Thus I do not believe that a "desire for comprehension" is
> the father of philosophy, but rather that a quite different desire
> has here as elsewhere used comprehension (together with mis-
> comprehension) as tools to serve its own ends. . . . Conversely,
> there is nothing impersonal whatsoever in the philosopher. And
> particularly his morality testifies decidedly and decisively as to
> *who he is* — that is, what order of rank the innermost desires of his
> nature occupy (*BGE*, 6, 7).

Nietzsche began his Preface to *Beyond Good and Evil* with these words:

> Supposing that Truth is a woman — well, now, is there not some foundation for suspecting that all philosophers, insofar as they were dogmatists, have not known how to handle women? That the gruesomee earnestness, the left-handed obtrusiveness, with which they have usually approached Truth have been unskilled and unseemly methods for prejudicing a woman (of all people) in their favor? One thing is certain: she has not been so prejudiced. Today, every sort of dogmatism occupies a dismayed and discouraged position — if, indeed, it has maintained any position at all. For there are scoffers who maintain that dogma has collapsed, even worse, that it is laboring to draw its last breath. Seriously speaking, there are good grounds for hoping that all philosophic dogmatizing, however solemn, however final and ultimate it has pretended to be, may after all have been merely a noble child's-play and mere beginning. And perhaps the time is very near when we shall again and again comprehend *how* flimsy the cornerstone has been upon which the dogmatists have hitherto built their sublime and absolute philosophical edifices. . . . The philosophy of the dogmatists, we hope, was only a promise held out over the millenniums, similar to astrology in still earlier times — astrology in whose service perhaps more labor, more money, more sharp-wittedness and patience have been spent than on any real science so far. To astrology and its "ultra-mundane" claims in Asia and Egypt we owe the grand style in architecture. It seems that in order to inscribe themselves upon the heart of humanity with everlasting claims, all great things must first sweep the earth disguised as enormous and fearsome grotesques. Dogmatic philosophy has been such a grotesque ... (*BGE*, xi, xii).

Aristotle's philosophy certainly qualifies as "great" and as "dogmatic." He has probably influenced Western thinking more than any single philosopher. His seminal treatises on metaphysics, ethics, politics, rhetoric, poetry, logic, biology, and psychology place him as perhaps the leading figure in intellectual history. However, in Nietzsche's interpretation Aristotle was a philosopher who had "not known how to handle woman," and whose philosophy, especially his morality testified "decidedly and decisively as to *who he is* — that is, what order of rank the innermost desires of his nature occupy."

Aristotle created an "order of rank" specifying the male as superior and dominant and the female as the opposite — inferior and subordinate. Naturally, that is — and universally. Nietzsche reminded us that value-ranking had for many centuries been the privilege — the power and prerogative — of philosophers or theologians. And they had invented interpretations of nature and of culture which placed them- selves at the pinnacle of the hierarchy of persons of value, and therefore of power. Nietzsche set his task as exposing these individuals — espe- cially Christian theologians and the religious hierarchy of values. He spared nothing and no one, as I attempted to reveal in *Nietzsche, Philoso- pher of the Perilous Perhaps*.

Nietzsche's critique of Christianity was essentially a critique of Christian values — as he understood them. *Nature*, the natural world — in all of its majesty, wonder, beauty, awesome powers — had been devalued in favor of an invented, false, other world. *Life*, in this natural world had been devalued, despised, denigrated and replaced in value with an imaginary, eternal, perfect life in this imaginary, invented world. *Sexuality*, the human *body*, male and female — had been devalued and replaced with an invented eternal soul. And ultimately, the female, and particularly the female body, had been devalued in favor of the male body — in this world or in the fabricated world. These words strike at the heart of Nietzsche's disgust:

> Reality has been deprived of its value, its meaning, its veracity to the same degree as an ideal world has been *fabricated*. . .The "real world" and the "apparent world" — in plain terms: the *fabricated* world and reality. . . The *lie* of the ideal has hitherto been the curse on reality, through it mankind itself has become mendacious and false down to its deepest instincts — to the point of worshipping the *inverse* values to those which alone could guarantee it pros- perity, future, the exalted *right* to a future (*EH*, 34).

> That contempt has been taught for the primary instincts of life: that a "soul", a "spirit" has been *lyingly invented* in order to destroy the body; that one teaches that there is something unclean in the precondition of life, sexuality; (*EH*, 132).

> I was the first to take seriously, for the understanding of the older, the still rich even overflowing Hellenic instinct, that won- derful phenomenon which bears the name of Dionysus: it is ex- plicable only in terms of an excess of force. . . . For it is only in the Dionysian mysteries, in the psychology of the Dionysian state, that the *basic fact* of the Hellenic instinct finds expression — its "will to life." What was it that the Hellene guaranteed himself by means of these mysteries? *Eternal* life, the eternal return of life; the future promised and hallowed in the past; the triumphant Yes to life beyond all death and change; *true* life as the over-all continuation of life through procreation, through the mysteries

of sexuality. For the Greeks the *sexual* symbol was therefore the venerable symbol par excellence, the real profundity in the whole of ancient piety. Every single element in the act of procreation, of pregnancy, and of birth aroused the highest and most solemn feelings. In the doctrine of the mysteries, *pain* is pronounced holy: the pangs of the woman giving birth hallow all pain; all becoming and growing — all that guarantees a future — involves pain. That there may be the eternal joy of creating, that the will to life may eternally affirm itself, the agony of the woman giving birth *must* also be there eternally.

All this is meant by the word Dionysus: I know of no higher symbolism than this Greek symbolism of the Dionysian festivals. Here the most profound instinct of life, that directed toward the future of life, the eternity of life, is experienced religiously — and the way to life, procreation, as the *holy* way. It was Christianity, with its *ressentiment* against life at the bottom of its heart, which first made something unclean of sexuality: it threw *filth* on the origin, on the presupposition of our life (*PN*, 560, 561, 562).

Just as most of the pre-Socratic philosophers saw the universe as an interplay of opposing forces, Bachofen recognized that the fundamental duality of existence was observed in the opposition of male and female, sun and moon, day and night, heaven and earth, and many others. One needs to read Bachofen's selected writings in the book, *Myth, Religion, & Mother Right*, to be persuaded (as I am persuaded) that much of Nietzsche's thinking developed out of the guiding influence of this Swiss philologist and cultural historian.

Nietzsche was familiar with the history of philosophical thought back to the classical philosophers, including of course, Aristotle, and to the pre-Socratics, especially Heraclitus. But Bachofen, I would claim, directed Nietzsche's attention to a much earlier time in human history and human development. Certain major themes in Nietzsche have a strong, even uncanny, similarity to those of Bachofen. In his early *Birth of Tragedy*, Nietzsche took over Bachofen's notions of the Dionysian and the Apollonian. He followed Bachofen in the recognition and affirmation of the power of religion as the dominant cultural determining force. He reaffirmed, if obliquely, the fundamental, original, natural contraries — male and female. And he clearly embraced Bachofen's interpretation that sexuality, the dynamics of the female/male relationship was, and had been, as far back into history as we have probed, the dominant theme in religion and culture. For both Bachofen and Nietzsche, it was nature (as well as culture, of course), the process of life and *all* which that entails, that captivated both of these devoted scholars. The following passage from Bachofen needs to be recalled:

> The relationship which stands at the origin of all culture, of every virtue, of every nobler aspect of existence, is that between mother and child; it operates in a world of violence as the divine principle of love, of union, of peace. Raising her young, the woman learns earlier than the man to extend her loving care beyond the limits of the ego to another creature, and to direct whatever gift of invention she possesses to the preservation and improvement of this other's existence. Woman at this stage is the repository of all culture, of all benevolence, of all devotion, of all concerns for the living and grief for the dead (MR, 79).

It is of interest that in his book, *Human, All Too Human*, published not too many years after his Basel experience, Nietzsche has a short tantalizing section called "Woman and Child." I would suggest that this topic probably would have been seen by his philosophical colleagues as not qualifying as a "typical" philosophical subject, as "unseemly." Recall this also from Bachofen:

> The progress from the maternal to the paternal conception of man forms the most important turning point in the history of the relations between the sexes. . . . The mother's connection with the child is based on a material relationship, it is accessible to sense perception and remains always a natural truth. But the father as begetter presents an entirely different aspect. Standing in no visible relation to the child, he can never, even in the marital relation, cast off a certain fictive character. Belonging to the offspring only through the mediation of the mother, he always appears as the remoter potency (MR, 109).

Bachofen was not one of those philosophical dogmatists whom Nietzsche had exposed. Like Nietzsche, Bachofen considered that he had laid an important foundation upon which others might build. The two did have important differences. Bachofen's theory of the evolution of culture ultimately interpreted the matriarchal society, out of which modern patriarchal societies had evolved, as the lower stage of development. The higher stage in Bachofen's interpretation was, of course, patriarchy, and he saw this evolution as progress, a positive evolutionary gain. Nietzsche, it may be claimed saw the opposite, and developed his theory of Christianity and the West as decidedly a loss and a decline from earlier ancient forms. However, the more one explores the

thinking of these two, the more convincing is the evidence that suggests a huge debt by Nietzsche to Bachofen. Recall this short passage
of Nietzsche's:

> This is why a philosopher should consider himself justified in in-
> cluding willing within the general sphere of morality — morality
> understood as the doctrine of the rank-relations that produce the
> phenomenon we call "life"... (*BGE*, 22).

We need to gather all possible resources to continue this task —
presumptuous as it may be, or appear to be — of expanding our understanding of *war* and of hoping, and believing, that Nietzsche's thinking
may hold some clues as to the direction to be taken. Questions have
been raised and challenges made to the uses and possible meanings in
Nietzsche's writings of the term "life." In further preparation for ap-
proaching the puzzle of war; for daring to presume that had Nietzsche's vibrant intellectual life extended beyond his forty-fourth year;
and, for further daring to attempt to put together what an interpretation of war by him might, or would, have been — I want to expand on
the significance of what I believe "life" meant for him. Philosophically
and personally.

Life is a *process*, or perhaps preferable, *living* is a process. Or, there is
a process called "living." And, process may be understood as "a natural
phenomenon marked by gradual changes," or "a natural continuing activity or function," or as "something going on." And, a *phenomenon* may
be understood as "an observable fact or event, or as "an object or aspect
known through senses rather than by thought or intuition," or as "a
temporal or spatio-temporal object of sensory experience." And further, *experience* may be understood as "direct observation of, or participation in, events as a basis for knowledge, or as "the process of directly
perceiving events or reality." Life, living — this process in which we are
constantly participating.

This *process* which is called *living* is simultaneously a natural phenomenon and a cultural phenomenon. Perhaps they could be considered as a fundamental pair of contraries — of a special type. Our observations, our perspectives, our interpretations, etc., may focus on one or
the other, but both are always the context in which our participating,
our living occurs. Perhaps like the yin/yang symbol. Nietzsche does
refer often to nature and to culture. Consideration of these separate

aspects, or "states," is fundamental in all of his thinking. As with his uses and possible meanings of the term "life," there has been similar questioning regarding the terms "nature" and "culture." Stepping into that murky stream, I only suggest this. *Nature*, or natural, seems to be referring to something like the inherent character of the universe; or the original condition of humans; or the initial, germinal stage of all processes. *Culture*, or cultural, seems to be referring to the integrated pattern of human knowledge, belief, and behavior that depends on man's capacity for learning and transmitting knowledge to succeeding generations; or, the customary beliefs, social forms, and material traits of a racial, religious, or social group; or, the set of shared attitudes, values, goals, and practices that characterize a group of people. Both Bachofen and Nietzsche, I would suggest, view culture as having derived, and is deriving, out of that initial germinal stage which is nature.

It may prove helpful later to elaborate here a little further on the general process called living — in Nietzsche's perspective, or perspectives. There can be little serious doubting that he favored the attitude, as he saw it, of the ancients, to nature and to life — one of a "lavish abundance of gratitude." He was the ultimate "life affirmer." So, indulge in a few reminders and a few particular processes which, in Nietzsche's world, are characteristic of life, or living. Life is constantly changing, is dualistic, agonistic, striving for the feeling of power, cyclic, linear, rhythmic, eternally recurring. Particular and special aspects which were of importance to Nietzsche include: creating, destroying, procreating, producing, possessing, dispossessing, perceiving, sensing, reading, interpreting, writing, speaking, communicating, thinking, learning, unlearning, feeling, benefiting, harming, questioning, answering, choosing, naming, defining, desiring, competing, hoping, winning, losing, remembering, forgetting, arriving, departing, understanding, misunderstanding, deceiving, concealing, revealing, doubting, challenging, believing, breathing, sleeping, smelling, listening, tasting, eating, drinking, singing, walking, running, playing, dancing, celebrating, laughing, imagining, symbolizing, and valuing. And so endlessly.

In Nietzsche's eyes, nature and life are rich, occurring in such abundance as to be beyond any comprehension. Both are complex and simple, harsh and gentle. And from his perspective, we may value nature and life or we may devalue both. Christianity, he insisted, had devalued both — at our own peril. Of all the activities, the processes, that make up living for human beings, I venture that creating and valuing are at the top of Nietzsche's list.

Nietzsche took as his task — his "destiny" — describing, dissecting, interpreting, evaluating, the destructive power of Christianity in the latter half of the 19th century. And, he elaborates extensively on his conclusions. During that time he also was constructing an image of the future, and of his involvement with that future. Here are a few of his final assessments:

> I know my fate. One day there will be associated with my name the recollection of something frightful — of a crisis like no other before on earth, of the profoundest collision of conscience, a decision evoked *against* everything that until then had been believed in, demanded, sanctified. I am not a man I am dynamite. . . .I was the first to *discover* the truth, in that I was the first to sense — *smell* — the lie as lie. . . My genius is in my nostrils . . . I contradict as has never been contradicted and am nonetheless the opposite of a negative spirit. I am a *bringer of good tidings* such as there has never been, I know tasks from such a height that any conception of them has hitherto been lacking; only after me is it possible to hope again. With all that I am necessarily a man of fatality. For when truth steps into battle with the lie of millennia we shall have convulsions, an earthquake spasm, a transposition of valley and mountain such as has never been dreamed of. The concept of politics has then become completely absorbed into a war of spirits, all the power-structures of the old society have been blown into the air — they one and all reposed on the lie: there will be wars such as there have never yet been on earth. Only after me will there be *grand politics* on earth (*EH*, 126, 127).

Nietzsche's "future" has become, or better, is still becoming our present. We must wonder, even doubt, whether he could have comprehended the extent of the earthquake, the convulsions, which would erupt beginning early in the 20th century, and gaining in strength in the 21st century. The first World War exploded into existence less than two decades after his death. He understood and had cited the centrality of *power* and *values* in the catastrophic decomposing of Western culture. The crises occurring now are crises involving losses of belief, of identity, of power, of values. There is rapid fragmenting of power within, and at the top of hierarchies of power, e.g., in religion, in politics, in the economy, in the military, in the family. Humans have invented the instruments of power sufficient to destroy most of our own species, or perhaps all life on our planet. It appears that we have adopted the Machiavellian "virtues" of fear, deceit, and cruelty in pursuit of power.

Nietzsche clearly believed that at his time the power and function of the Christian religion had to be the main subject of critical thinking. The further demand now is for a similar thinking, or re-thinking, of war. The imperative is to think war "to a greater depth," to raise new, difficult, dangerous questions, to appeal to different perspectives, to develop new interpretations — to engage in a critique of war with Nietzsche leading the way.

Earlier in this book, in looking for perspectives which might serve as suggestive points or clues as to how we might take the first steps in the daunting and presumptuous task of "thinking war to a greater depth," it seemed persuasive, historically, to hear from the Greeks — the poets, Hesiod and Homer, the pre-Socratic philosophers, the dramatists, and the historians, Thucydides and Herodotus. We went back to the 7th or 8th centuries BC. Before listening to Nietzsche, there is a very recent perspective and interpretation which I found advanced immeasurably and profoundly any understanding we may have of war.

Drew Gilpin Faust, President of Harvard University, in her latest book *This Republic of Suffering*, has written a remarkable, highly-praised, powerful account of the American Civil War. A first look at the chapter titles in the table of contents shows the reader that what lies ahead is not the conventional, traditional approach to war — to this particular war. We have words about war from a different perspective, spoken in a different voice — that of an eminent historian, a highly respected and successful woman. The chapter headings — "Dying," "Killing," "Burying," "Naming," "Realizing," "Believing and Doubting," "Accounting," "Numbering," and an Epilogue "Surviving," speak vividly and uncompromisingly what this war — and by extension every war — is *about*. On the jacket of her book, there are these words, "Were he alive today *This Republic of Suffering* would compel Walt Whitman to abandon his certainty that the 'real war will never get into the books'." Faust's recounting of this particular civil war (the kind of war seen by Hobbes as the worst type) says to me that war is a process, one with multiple elements. It is about "what goes on" during this process — the "real war," and the real persons engaged and affected. Faust personalizes war.

And finally to Nietzsche. The process of persuading him to speak concerning war is basically this. To restate again, and reemphasize, some things which he did say, which he himself emphasized, around which his entire philosophy was continually being shaped and re-shaped — until his time ran out. From these sources, our interest is not to call attention to what he did *not* say. Rather, the process is to think through what he could have said, might have said, even daring to consider what he would have said, to have replied to our question, "Why War?" It is a challenge which I am convinced Nietzsche would have applauded and in which he would have been, or would be now, an enthusiastic participant.

Using broad strokes, we are able to sketch this background. Nietzsche was among a few others in rejecting philosophical meta-physics, theology, abstract and dogmatic philosophy. He welcomed the rapidly growing influence of science — on new questions of method, of perception, of the senses, of the excitement of new possibilities, new frontiers of knowing, of thinking. He rejected the idea of any single, common, "right" perception, declaring instead that all perceiving is a matter of perspective, is "perspectival," personal, open, unlimited. And, he insisted that all knowledge, all knowing, is a matter of interpreta-tion, of assigning meaning. And that, as with perspective, interpreta-tions are initially and basically unlimited. He enthusiastically adopted the idea of experimenting, theorizing, verifying, falsifying.

So many new sciences, developing so rapidly. It would be fruitful to argue that this was the most significant factor in the intellectual envi-ronment to influence Nietzsche. From among these emerging sciences, those which attracted Nietzsche the most were philology, physiology, psychology, and history. He became a scientific philosopher, a physi-ological psychologist, and a cultural historian and critic.

Central to the rejection of abstract, dogmatic philosophy and the-ology, was the denial of any view of a stable, fixed, objective, perma-nent universe. Essential to any scientific interpretation was the idea of *evolution* — the process of continuous change. This became the basic principle of all scientific thinking and certainly for Nietzsche's think-

ing. The universe, nature, life, the body, thinking, language, power relations, consciousness, values — everything, in his view and words, was
in a state of "becoming" — moving, changing, in process.

At least two additional major cultural processes had been taking
place. One was the rediscovering of nature, of the human body, especially in the visual arts and in literature, and then in the sciences
— medical science, biology, anatomy, and physiology. And also for
Nietzsche, in Bachofen's mythology. We may add here that there is
little disagreement among major interpreters of Nietzsche in interpreting his writings as his ceaseless effort to redeem nature, natural life, the
body — to reposition man firmly back into his natural world, and to
reaffirm, reclaim, and revalue this world once again. The long separa-
tion of man from the initial germinative environment and the processes
of nature and life had proven to be a natural and cultural disaster.

The second cultural process was the emerging of the so-called man/
woman problem, what I earlier referred to as "gender-consciousness." In
1792 the English writer, Mary Wollstonecraft, published *A Vindication of
the Rights of Woman*, in response to the dominant, centuries-old, continuing
views, both philosophical and popular, that women were naturally inferior, rationally deficient, inherently lacking in the defining aspect of man
— reason. Wollstonecraft was urging equal rights for women, especially
in education and training. She referred to "The preposterous distinctions
of rank, which renders civilization a curse, . . ." Wollstonecraft brought a
new and distinctive voice to the revolutionary fervor that was sweeping
across Europe.

We are trying to find our way to Nietzsche's unfinished way with
war, or wars. Consider his emphatic insistence that philosophy, to be
relevant or successful, must remake itself in the image of new emerging
scientific and historical perspectives. He himself became a cultural historian and cultural critic — encouraged to a large extent by the studies
of Bachofen and his contributions to our understanding of the irreducible significance of Greek and Roman mythology and symbology. And
the two defining issues that Bachofen lays bare — *sex* and *power*.

If, as I believe, the so-called man/woman problem, or puzzle, was
central during Nietzsche's entire life and career; if the issues of power
and sexuality had been subjects for thinkers for centuries prior to the

18th and 19th centuries; and, if sexuality and sexual politics after Nietzsche's time, and continuing into the present, has become the formidable concern that it appears to be — then we will adopt an historical perspective. It may appear to be a diversionary or distracting enterprise, but I believe it is important, perhaps imperative, for the terrain toward which we are moving.

In the earlier decades of the second half of the 20th century, there was emerging a wealth of new creative thinking and writing from both women and men on the subject of woman and man. Before that time such ideas and writing had been products almost exclusively of males. A book entitled *History of Ideas on Woman*, by Rosemary Agonito, was published in 1977 by G. P. Putnam's Sons. The book is recommended on the cover as follows:

> Here, for the first time in a single volume, are the primary sources of Western Civilization's attitudes toward women. From Genesis and Plato, to Rousseau and Hegel, Engels and Freud, these writings are the key thoughts that have shaped our concept of women's physiology, psychology, legal rights, virtues, vices, origins and ethics (*IW*, cover).

Out of twenty-eight writers included in Agonito's book, twenty-four are male, four (including Wollstonecraft) are female. Note here that Nietzsche recognized and commented on the obvious fact that men had created the images of women, and women, in creating themselves according to these images, had become corrupted. More than any of the other writers, Aristotle in the 4th century BC and Thomas Aquinas in the 13th century, relying heavily on Aristotle's views, were the main movers and shapers of Western culture. Philosophical and theological eminences were a formidable alliance. The Greek and the Judeo-Christian traditions were skillfully combined. A third philosopher, Arthur Schopenhauer, immediately preceding Nietzsche, also played a major role in influencing Nietzsche's response.

Brief references to these three — Aristotle, Thomas Aquinas, and Schopenhauer — are intended to add further to the notion, or question, of how inextricably the man/woman problem is related to any understanding of war. And I have chosen to use the brief accurate summaries provided by Agonito rather than piece together individual direct quotations from the three. Regarding Aristotle:

Of particular interest are his biological writings, especially "The Generation of Animals," which investigates the reproduction of life and the female's role in this, and his political work, *Politics*, which discusses woman's status in society. Although the dates of these works are disputed, they are believed to have been composed between 347 and 322 BC In essence Aristotle maintains that woman is a mutilated or incomplete man, a thesis that has enjoyed a long and persistent history culminating in Freud's theories about woman. Aristotle accepts the pre-Socratic notion that women are colder than men. Since he associates heat with life or soul, he therefore supposes women to have less soul than men. Moreover, because woman is deficient in vital heat, she is unable to "cook" to the point of purity her secretion (menstrual blood), which unites with a male's secretion (semen) to form a new human being. The male's contribution to generation is pure and hence more valuable than woman's, although both, of course, are necessary. Specifically, man contributes the form or essence of the embryo while woman merely provides the nutrition necessary to maintain it; that is, the male, not the female, creates human life. Ultimately, woman's deficiency, or lack of the male principle — sentient and rational soul — affect the bodily, intellectual, and social status of women for the worse. She is physically weaker, less capable of rational thought, and subordinate to the rule of men (*IW*, 41, 42).

Then there is Thomas Aquinas:

By the thirteenth century the medieval Church had long since solidified its thinking about women, following the lead of Paul and Augustine. *Summa Theologica*, the multivolume work Aquinas wrote between the years 1266 and 1272, reflects official attitudes. By Aquinas' own account the *Summa* was conceived as providing a systematic treatment of the theological doctrines inherent in Scripture. He is concerned to present rational justification for these doctrines wherever possible and so relies heavily on the philosophical tradition dating from the Greeks, and in particular on Aristotle, who he felt had discovered universal truths. Aquinas' reflections on women, therefore, like much in the *Summa*, skillfully combine the Judeo-Christian and Greek traditions. For example, the Christian acceptance of the subordinate status of women finds expression in the Aristotelian idea that this is a natural state owing to man's greater rationality. Aquinas accepts Aristotle's notion that woman is a defective male lacking vital force as true of each woman, although from a Christian point of view, since God created woman, the creation of woman in general is not tainted by defect but is natural and ordered like every divine act. In addition, the Judeo-Christian attitude to man's dignity requiring the human race to descend from man, not woman, finds its parallel in Aristotle's view that the male supplies the essentially human features in reproduction while woman merely provides the nutritive material. Aquinas' *Summa Theologica*, including its treatment of women, was destined to provide the Roman Catholic Church with its official theological and philosophical dogma for centuries to come (*IW*, 81, 82).

And then immediately preceding Nietzsche, Arthur Schopenhauer:

Schopenhauer's essay, "On Women" (1851) has come to be known among twentieth-century feminists as a classic of mysogynist writing. He combines virtually all the negative aspects of traditional ideas on womankind and adds one or two of his own. Schopenhauer's basic contention is twofold — first, that women's qualities (their behavior, virtues, vices, etc.) are natural and second, that by nature women are in all respects inferior to men. This inferiority is qualitative (men have certain qualities that women lack) and quantitative (women are deficient in the qualities they share with men). Specifically, women are deficient in reason, in physical strength, and in love, — and a new twist — in the aesthetic faculty. All these flaws have serious consequences. Woman's deficient intelligence accounts for her lifelong childishness, her willfulness, and her lack of a sense of justice, since justice depends on deliberation, and explains why woman must obey and man rule by nature. She is not physically strong, and so nature makes her cunning, a master of pretence and lying. The natural cunningness, in turn, explains why women are by nature false, unfaithful, hypocritical ingrates, among other things. Woman's deficiency in love makes her hate her own sex and unable to love her own children on any level other than the purely instinctive. She is capable only of loving man, since this is her whole goal in life. Finally, she has no aesthetic faculty (which goes with her naturally ugly appearance) and so cannot appreciate any fine art. In short, to reverence woman in a chivalric spirit is to flout nature at every turn (*IW*, 191, 192).

Later writers such as Ralph Waldo Emerson (a favorite of Nietzsche's), John Stuart Mill, and Ashley Montague were kinder and gentler.

Another book, written about the same time as that referred to above, is entitled *Sex and Power in History*, by Amaury de Riencourt, published by Dell Publishing Co., Inc., in 1974. Regarding this widely researched and scholarly book, commentary on the cover reads:

In this unique sexual interpretation of history, Amaury de Riencourt traces the changing historic roles of women and reaches back to remote sources to explain — and suggest solutions to — the current predicament of the sexes. When some three or four thousand years ago men revolted against the myth of the Great Earth Goddess and set up dominant male gods, it was a psychological event of the first magnitude. This shift to the mas-

culine principle, claims de Riencourt, marked the beginning of
history proper and led to the male-oriented societies of Greece
and Rome. From Rome's fall, de Riencourt follows the fortunes
of women from tribal and feudal to Renaissance and industrial
societies and makes some startling predictions about the direction of Western human ecology (SP, cover).

A few excerpts from de Riencourt's detailed history of the female/
male relationship and the dynamics of sexual politics through history affirm the impact, especially of Aristotle, Thomas Aquinas, and
Schopenhauer on that history. These references also help to tighten up
a number of the major influences within which Nietzsche astonished
the world. These from the Introduction:

> Essentially, this work is an all-inclusive interpretation of history
> from end to end, as seen through the interplay of the primary bio-
> social forces that have shaped it — the *yin* and *yang*, the female
> and male principles that are the warp and woof of an intricate
> tapestry. This interplay underlies the natural evolution of practi-
> cally all living organisms, including humankind. But the *yin-yang*
> dialectic also permeates the historical development of all cul-
> tures and civilizations, all mythologies, religions, philosophies,
> arts, political and social organizations, economic doctrines, and
> structures.
>
> Only the sick feel their limbs and organs. The acute conscious-
> ness of a predicament in the relations between the sexes is prob-
> ably the most significant element in the overall crisis of contem-
> porary civilization, because it subsumes all the others. What
> follows is an endeavor to analyze this crisis by going back to its
> remote sources and giving a coherent account of the weaving of
> this planetary fabric: the human race (*SP*, vii).
>
> At some point, however, men gradually substituted a male god
> for the former Mother Goddess when they discovered the con-
> nection between sex and procreation, and their *biological* pater-
> nity. This *patriarchal revolution*, which swept the whole world
> some thirty-five hundred years ago, was preceded in the collec-
> tive unconscious by the mythological metamorphosis known as
> *solarization*, the victory of the male sun god over the female moon
> goddess. In turn this implied the collapse of the female-oriented
> *cyclical* fertility cults and the rise to supremacy of the male con-
> cept of *linear* history consisting of unrepeatable events, messian-
> ism, and eschatology, with Zoroaster and the Hebrew prophets.
> Alongside of this metaphysical transformation, Greek rational
> thought broke away from magic thought processes, and sym-
> bolically transferred the center of creative power from the female
> womb to the male brain (symbolized by Athene's birth from the
> forehead of Zeus), creating the great Western cultural distortion

that is still with us today, by giving greater value to culture than to nature, to the abstract *Idea* than to concrete *Life* itself. This entailed that, while the masculine Promethean drive was exalted, while Greek thinkers became the pioneers of philosophy and discursive thought, the *psychological* degradation of the female principle (physiological creation of new life) led to the debasement of woman's *social* position and status, which reached its nadir at the time of Pericles. This cultural distortion — unknown to Eastern civilization where a somewhat harmonious balance between the two *sexual principles* was maintained — disrupted the Western human ecology by depicting females as inferior or *incomplete* males (biblical Eve, Greek Pandora, and Aristotle's philosophic rationalizations) instead of granting them their *different* specificity, as was done in China or India (*SP*, viii).

The *cultural* devaluation of woman's life-creating function was thus compounded by a *social* deactivation of bourgeois women as economic producers, as if they were obsolete machinery. This gave further impetus to the feminist revolt, which if not understood and coped with at its *cultural root*, could destroy Western society as a similar movement destroyed Roman civilization (*SP*, ix).

Social engineering and legislative tinkering will not solve the problem. The present dilemma can be overcome only if a profound *cultural* change takes place and an entirely new set of values replaces the traditional values that Western civilization inherited from both biblical and Greek patriarchal sources. These new values will be effective only to the extent that they are based on respect for the different specificities of the sexes and the same reverence of Life as for the creations of the Mind (*SP*, x).

In examining the Medieval Period de Riencourt writes:

In spite of chivalry and the increasingly popular cult of the Virgin, theology, relying mostly on Aristotle, took a dim view of the female sex. Johannes Scotus Erigena, that ninth-century Irish beacon shining in the Dark Ages, had already propounded the thesis that originally mankind was sinless and therefore without distinction of sex; it was as a result of sin that human beings were divided into males and females, the female embodying mankind's sensual and sinful nature. For Thomas Aquinas woman was an inferior human being, defective and accidental (deficiens et occasionatum), a male that was askew, the result of a weakness in the procreative powers of the father. Aquinas remained faithful to the old concept of classical Greece (as interpreted by Aristotle) that woman provides only inert *matter* to her offspring while man provides the spiritual entity, the active *form*. Woman is mainly a sexual appetite, while man is better balanced; weaker in every respect — in will, mind, and body — she is to man what the physical senses are to reason. Her sole purpose is procreation, since man can do everything better than she does — even her domestic chores. Unable to occupy any important position in either church or state, she is meant to look up to man as her natural lord and accept his authority without question. (*SP*, 227)

These are some of de Riencourt's words near the end of the book:

The greatest error made by Western man since the days of Aristotle, repeated and emphasized by Thomas Aquinas and many others, is precisely to slur over the "difference" and mistakenly see woman as a defective, incomplete, lower-grade male who *lacks* something — *un homme manqué*; to see what is essentially the other as basically the same, but misbegotten and of lesser quality. As Thomas Browne expressed this traditional view: "Man is the whole World, and the breath of God; Woman the Rib and crooked piece of man." This mistake was never made in civilizations such as the Indian and the Chinese where the difference never implied ultimate superiority or inferiority. If Western woman now attempts to become a pseudo-male, it is largely because Western man invented this concept long ago, instead of granting female specificity the full recognition and respect it deserves — result of an overvaluation of cerebral creation at the expense of the physiological maternal creativity that gives birth to it (*SP*, 422, 423).

From the Modern Age these are some of the words of de Riencourt:

With Schopenhauer, whose doctrine of the primacy of the will was in direct conflict with Hegel's idealism, we reach a stage of antifeminist bitterness that goes well beyond the usual Teutonic diatribes against the second sex — a bitterness that heralds the dawn of a new age in terms of a direct conflict between the sexes that would have been meaningless in preceding centuries. In his *Studies in Pessimism*, he berates "our old French notions of gallantry, and our preposterous system of reverence — the highest product of Teutonic–Christian stupidity. These notions serve only to make women more arrogant and overbearing; so that one is occasionally reminded of the holy ages in Benares, who in the consciousness of their sanctity and inviolable position think they can do exactly as they please" (*SP*, 291, 292).

German abstract thought has dominated the nineteenth and twentieth centuries. German philosophy that ranges from Kant to Hegel to Schopenhauer and Nietzsche has no equivalent in other Western lands. Distinctly German also are the thought processes of Marx, Engels, Freud, Jung, and Einstein who, in their respective fields, revolutionized the world. And German pathos has given the world its greatest music from Bach to Wagner and beyond. All the great intellectual and artistic productions of the Germans, as well as their pharaonic distortions in the political world, are the result of this uneasy marriage within the bisexual German male's soul between the masculine and feminine principles (*PS*, 292, 293).

Schopenhauer's typical misogyny becomes more understandable against this background. His violent attack against the female sex came a few decades after the first feminist blast sounded in England. Mary Wollstonecraft's pioneering work, *Vindication of the Rights of Women* (1792), was the first clarion call for the forthcoming battle of the sexes, the first appearance and explicit statement of a problem that had not existed previously in Western consciousness — since the distant days of the Greco-Roman antiquity. . . Quite apart from her personal eccentricities, Mary Wollstonecraft's thoughts and activity implied a sense of profound change; they betrayed the fact that a revolutionary undercurrent was about to break into the open and violently upset the traditional relationship between the sexes in Western civilization (*PS*, 293, 294).

Referring to Nietzsche, de Riencourt says:

The last step in the reintroduction of the feminine principle at the heart of philosophic thinking was the reemergence of cyclical interpretations of history, contrasted with the traditional, strictly linear, eschatological, and masculine interpretation — to which Marxism's apocalyptic vision remains faithful in spite of its dialectical interpretation. One of the first portents of this new trend was Nietzsche's theory of "eternal recurrence," the metaphysical theory that humankind has been, is, and will forever be undergoing the same life course from beginning to end, thus rejecting simultaneously the Christian doctrine of an afterlife and the fashionable belief in endless progress that was inaugurated by the Enlightenment. For all his ranting against woman, Nietzsche was by far the most "feminine" of the nineteenth-century philosophers — (*PS*, 297, 298).

Nietzsche prophesied the inevitable doom of European culture, the collapse of all traditional values, and the triumph of nihilism, along with the triumph of the Dionysian spirit — the full acceptance and, eventually perhaps, the transcending of decadence. Today, we can see and feel the decadence, but certainly not the overcoming of it. The Western Promethean drive is coming to an end, and all its components are beginning to fall apart. Quite naturally, one of its most important components, the relatively harmonious relations between the sexes, is also falling apart; now that Prometheus's sons are faltering, Pandora's daughters revolt — not gently, in womanly fashion as they did some generations ago, but with cold, revolutionary fury, in the *masculine* mode (*PS*, 402).

And, in addressing the end of Hesiod's Iron Age, de Riencourt writes:

> The disintegration of Western culture, by itself, does not entail the end of the world; the human race can still live on and prosper under the aegis of other ideals and values. But what is coming to an end with it is a five-thousand-year historical cycle that started at the end of the Bronze Age with the patriarchal revolution. The reason for this is obvious — the far-reaching consequences of the technological revolution. Like wheels within wheels, the end of Western culture's life course is also coinciding with the end of all other separate cultures, with the end of the Iron Age and, hopefully, with the establishment of a global, humankind-wide civilization (PS, 419).

PART FOUR

Given his studies both in philosophy and in classical philology, Nietzsche's interests and endeavors in history became a principal element in his thinking. It would be difficult not to assume that his historical consciousness far exceeded that of any of his contemporaries — or even of those who followed after him. For Nietzsche, not history in the abstract or theoretical, but history as he was living it, thinking it, writing it. And naturally, living his present meant also learning and re-thinking the past, and racing ahead, predicting the future — separate periods, but forming a continuing process. He was acutely aware of his own moment on the stage of history. Nietzsche had the insight to understand what he believed his own age needed and desperately demanded. This was his "timeliness." His "untimeliness" was his recognition that his new ways of thinking were too dangerous, too revolutionary, to be heard. He must wait until a later time. Remember, he said, "Some are born posthumously."

Nietzsche was adamant in believing, and in attempting to convince others, that in order to remain, or to become relevant, philosophical thinking must itself become historical, as well as scientific. With that in mind, it seems appropriate and worthwhile at this juncture to sketch in broad strokes a somewhat different perspective of Nietzsche and history.

Beginning with Nietzsche's *present,* the latter half of the 19th century, the Zeitgeist — the general intellectual, moral, and cultural climate — as we have noted, appeared to be evolving rapidly in these important respects. There was increasingly a declining interest in, and rejecting of, abstract, dogmatic, metaphysical, supernatural, explanations or interpretations, both theological and philosophical. At the same time, the interest in the rapidly expanding possibilities offered by the emergence of Science — renewed emphasis on sense experience, on knowledge, on method, on evolution as a process of change — was staggering and revolutionary. New individual sciences were proliferating. An exciting emphasis on nature, on life, on the human body, was reversing the earlier centuries-old emphasis on the supernatural, on some future "other" life, on the human soul. This was becoming obvious in the arts, in literature, in the sciences, and in philosophy.

Into this volatile atmosphere there arrived an unforeseen, unexpected, incendiary, improvised, explosive device — Mary Wollstonecraft's *A Vindication of the Rights of Woman.* A different voice began to trigger unfamiliar sound waves which began to reverberate rapidly throughout the cultural fabric of Europe. What I have referred to as evolving "gender-consciousness" was inescapable, and the responses were of many types and degrees of intensity. It became for Nietzsche a major attraction.

And what of Nietzsche, to what he referred to as his "fatality," his "destiny"? Again, in broad strokes. He arrived in this maelstrom as a male, in Germany, in 1844. Or, as Heidegger might say "he was thrown into a world which was already going." His father and both grandfathers were Lutheran clergymen. At the age of twenty, like many of his peers, Nietzsche entered Bonn University as a student of theology and classical philology. He subsequently abandoned theology and discovered philosophy in first reading Schopenhauer. His renunciation of his religion and of the study of theology was a critical moment in his evolution. Having transferred to Leipzig, he became a classical scholar and was appointed to the chair of classical philology at Basel at the extraordinary age of twenty-four, where he was in frequent association with

Wagner and with Bachofen. His earlier renunciation of his religion was followed by his relinquishing his German nationality and applying for Swiss citizenship. Nietzsche *became* a philosopher. Of his philosophy, R. J. Hollingdale wrote this:

> Nietzsche's philosophy is not to be found in any single book, nor in the books of any single period of his life. It is, rather, an aggregate of all of his books. It is a process of development, not a body of propositions.
>
> To "know" his philosophy therefore amounts to experiencing this development. There exists a widespread belief that it is a matter of indifference which of his books you read first or in what order you read them: but the opposite is the case if your intention in reading them is to understand his philosophy, for his philosophy is, so to speak, his mind's autobiography of which each book is a chapter. You will get far more out of him if you regard the entire series of books as if it were *one* book (N, 43).

Classical philology, philosophy, mythology — assured his becoming "historical" — a cultural historian, and the most unremitting critic of his own German, and Western, culture.

Nietzsche was a philosopher of his own time, living his *present*. But also, at the same time, his attention was focused on living the *past*. Most, perhaps all, of the ideas which were becoming the richness of his philosophy had been threading their way in and out of the Western intellectual tradition since the early myth-makers and pre-Socratics. Power, values, nature, life, death, contraries, bodies, sexuality, and sexual politics (politics understood as "the use of intrigue and strategy to obtain a position of power").

All of this, plus the emergence of two new sciences: physiology and psychology. For Nietzsche, the necessary *interdependence* of these meant that he continued his philosophical development with the addition of what he referred to as a "physiological psychologist." Nietzsche *became* a "new" psychologist. It would be hard to exaggerate the impact this would increasingly have on shaping and reshaping his thinking. In *Beyond Good and Evil*, he wrote:

> All psychology hitherto has become stuck in moral prejudices and fears: none has ventured into the depths. To consider psychology as the morphology and evolutionary doctrine of the will to power — as I consider it — this no one has touched upon even in thought (insofar as it is allowable to recognize in what has been written the symptoms of what has been kept dark). The

force of moral prejudices has penetrated deeply into the most spiritual, the seemingly coldest and most open-minded world, and, as one may imagine, with harmful, obstructionist, blinding, and distorting results. A proper physio-psychology must battle with unconscious resistances in the heart of the investigator; his "heart" sides against it. Even a doctrine of the reciprocally limiting interaction of the "good" and "wicked" impulses causes, as being a subtle form of immorality, some distress and aversion in a still strong and hearty conscience. Even worse is a doctrine that all the good impulses are derived from the wicked ones. But imagine someone who takes the very passions — hatred, envy, greed, domineering — to be the passions upon which life is conditioned, as things which must be present in the total household of life. Takes them to be necessary in order to preserve the very nature of life, to be further developed if life is to be further developed! Such a man suffers from the inclination of his judgment as though from seasickness! But even this hypothesis is by no means the most painful or the strangest in this enormous, almost totally unknown domain of dangerous insights. Indeed, there are a hundred good reasons for staying away from it if one — can! On the other hand, if our ship has once taken us there — very well, let us go ahead, grit our teeth, open our eyes, grip the rudder and — ride out morality! Perhaps we will crush and destroy our own remaining morality, but what do we matter! Never yet has a *deeper* world of insight been opened to bold travellers and adventurers. And the psychologist who can make this sort of "sacrifice" (it is not the *sacrifizio dell'intelleto* — on the contrary!) will at least be in a position to demand that psychology be acknowledged once more as the mistress of the sciences, for whose service and preparation the other sciences exist. For psychology is now again the road to the basic problems (*BGE*, 26, 27).

This "new" science, pursued by this "new" psychologist, promised to open up a "new" unexplored world — the inner world of the psyche, as contrasted with the outer world. Nietzsche was aware that with the development, the discoveries and interpretations, offered by the new psychology, every aspect of thinking would be affected in unimagined, unforeseen, unpredictable ways.

Every issue, every problem, every situation, every idea must be perceived, interpreted, thought about, spoken about, through the specific lens of psychology, "mistress of the sciences." Those dominant ideas — power, nature, conflict, culture, life, bodies, sexuality, process, values, morality — and the interrelatedness or interaction among these ideas — must be scrutinized microscopically. Philology, philosophy, history, mythology opened new accesses to the *past*. Psychology, in Nietzsche's view, would supply exciting clues to the *future*. Old philosophical approaches to understanding the human psyche, human nature, human behavior (which he refers to as "the thick-witted psychology of former times") were inadequate and outdated. And Nietzsche embraced the challenges wholeheartedly and completely — even if with trepidation. R. J. Hollingdale, more than any other translator and interpreter of Nietzsche of whom I am aware, addresses certain psychological phenomena which were "discovered" by Nietzsche and which are important for him and for us. With the counseling of Hollingdale in his excellent book, *Nietzsche*, and revisiting Nietzsche himself, the following highlights are significant.

Once Nietzsche had become convinced that what he called "will to power" was the defining factor of life, especially of human life, this was the center around which *all* psychic phenomena revolved. Human nature, all human behavior, had its origin in this drive, this impulse, this single basic motive.

Among a number of ways of addressing the various activities of Nietzsche which can be termed psychological investigations, Hollingdale refers to him as a "Psychologist of culture" and says this:

> What I mean by this phrase is an investigator of the psychological conditions underlying the phenomena of religion, art, philosophy and the other "higher" activities of man. . . . To seek the psychological origin of a "cultural fact" was almost an instinct with him, and it produced some of his finest and most characteristic writing (N, 178).

We know that probing for the psychological origin of religion became the basis of Nietzsche's devastating critical interpretation of Christianity. Taking a broader definition of culture — e.g., "the set of shared attitudes, beliefs, values, goals, social forms, and practices that characterize a group of people" — I am convinced, would have drawn Nietzsche's attention to the investigation of the psychological origin of violent warfare. And as with most of his thinking, these investigations would have resulted in rejecting the outdated and inadequate explanations of the past. We should not be surprised by new perspectives, broader, deeper, perhaps shocking interpretations. He would surely have been ready.

In addressing Nietzsche's life as having "consisted in living in solitude and opposing the current ideas of his time," Hollingdale wrote this:

> Such an isolated and remote figure is clearly not going to consider it his purpose in life to hold down a chair of philosophy or, indeed, a job of any kind: his function is to be "the man of the most comprehensive responsibility who has the conscience for the collective evolution of mankind" (*BGE*, 61), and as such he is not to be confused with "the philosophical labourers and scholars in general" (*ibid.*, 211). This distinction appears to have been forced on Nietzsche by his conception of the world as having been deprived of values and become as it were masterless through the death of God. In one of the earliest sections of *Human, All Too Human* he had written that "since the belief that a God directs the fate of the world has disappeared. . . mankind itself must set up oecumenical goals embracing the entire earth"; and for such "universal rule" not to be self-destructive, mankind "must first of all attain to an unprecedented *knowledge of the preconditions of culture*". It is in this that there lies "the tremendous task facing the great spirits of the coming century" (*HA*, 25). A few years later the solitary philosopher is seen as the voice and conscience of mankind in this matter, and he is consequently not to be confused with anything else: "It may be required for the education of a philosopher that he himself has also once stood on all those steps on which his servants, the scientific labourers of philosophy, remain standing. . . all these are only preconditions of his task: this task itself demands something different — it demands that he *create values. . . Actual philosophers. . . are commanders and lawgivers. . .* it is they who determine the Wherefore and Whither of mankind" (*BGE*, 211). In case it should be thought that he is here describing himself, he adds the three questions: "Are there such philosophers today? Have there been such philosophers? *Must* there not be such philosophers? . . ." (*ibid.*) Such philosophers are philosophers of the future, just as *Beyond Good and Evil* itself is, as its subtitle says, a "prelude to a philosophy of the future". "All the sciences," Nietzsche says elsewhere, "have from now on to prepare the way for the future task of the philosopher: this task understood in the sense that the philosopher has to solve the *problem of values*, that he has to determine the *order of rank of values*" (*GM*, I, note), (*N*, 172, 173).

At the center of Nietzsche's philosophy is power, and in discussing, interpreting, and clarifying this fact, this idea, Hollingdale is the best. From him we have this:

> *Psychologist of will to power.* This has been discussed in an earlier chapter and need not be repeated. "Will to power" as the basic psychological drive in man, of which all other human activities

are sublimations, is Nietzsche's specific psychological "theory", developed in the series from *Human, All Too Human* to *The Gay Science*, stated in *Zarathustra*, and explained and developed in the series from *Beyond Good and Evil* to *Ecce Homo*. The view of Nietzsche which asserts that he was constantly changing his mind, or that his "philosophy" is a programme of continual self-reversal and oscillation between opposites, is really refuted by the evolution of this theory, whose content it is obliged to ignore.

What is desired, according to this theory, is the feeling of increased power. The negative aspect, that is to say the feeling of impotence, of being subject to the power of another, produces as its characteristic effect the phenomenon of *ressentiment* — and this is the chief corollary of the theory of will to power. Nietzsche developed the psychology of resentment almost as luxuriently as he did that of power: the essence of it is that the powerless man feels resentment against those whose power he feels and against this state of powerlessness itself and out of this feeling of resentment *takes revenge* — on other people or on life itself. The objective of the revenge is to get rid of the feeling of powerlessness: the forms it takes include all moralities in which punishment is a prominent feature; all doctrines of future damnation (of the powerful) and salvation (of the powerless); the socialist millennium, with the class at present oppressed on top and the present ruling class exterminated or in labour camps; anarchism, which Nietzsche saw as a peculiarly direct and uncomplicated expression of the resentment of impotence; in short, every kind of open or disguised revenge. At its simplest and most personal, resentment is the opposite of magnanimity. The magnanimous man is essentially the man whose feelings and actions are prompted by an inner power: he is a happy man — his magnanimity is a consequence of his happiness. The resentful man is the reverse of this: his feelings and actions are prompted by an inner impotence, and his resentment and the effects it produces are a consequence of his unhappiness. Here again you will find Nietzsche is consistent: *ressentiment*, as he terms it, and the products of *ressentiment*, are always bad, and he would like to see the world rid of them (N, 183, 184).

Nietzsche, a man of many words, but one who believed also in brevity and clarity, especially when the need was for a precise, memorable definition, wrote the following:

What is good? Everything that heightens the feeling of power in man, the will to power, power itself.

What is bad? Everything that is born of weakness.

What is happiness? The feeling that power is *growing*, that resistance is overcome (PN, 570).

For Nietzsche, understanding what it means to be human must begin with — "life is the will to power." This drive, or impulse, defines the essential, eternal nature of all life, including and especially human life. Ultimately, every activity, however it may appear otherwise, may be explained on the basis of this single, original, necessary motive. This will to power is the permanent center of every psyche. Nietzsche recognizes that there are strong wills and weak wills — neither free wills nor equal wills. In some cases there is a lust for power. My reading of Nietzsche suggests that *all* other psychological phenomena are constantly in motion, often in conflict, advancing and receding — but always interacting with this dominant drive for the *feeling of power*. And this striving will to power in itself is neither called good nor bad — it simply *is* what life *is*, and therefore a necessity.

In considering further the difficulties and challenges facing him as he was becoming increasingly convinced of the enormous impact of these newly interpreted psychological phenomena on our understanding of human nature, on human behavior, on human culture, Nietzsche found it necessary to reinterpret, or reconceptualize, the idea of a "psyche," to disassociate himself firmly from either the Greek connotation or that of Christianity, of "soul." In *Thus Spoke Zarathustra*, Nietzsche speaks unequivocally of the direction his thinking was taking, and would take in the future. He writes:

> I want to speak to the despisers of the body. I would not have them learn and teach differently, but merely say farewell to their own bodies — and thus become silent.
>
> "Body am I, and soul" — thus speaks the child. And why should one not speak like children?
>
> But the awakened and knowing say: body am I entirely, and nothing else; and soul is only a word for something about the body.
>
> The body is a great reason, a plurality with one sense, a war and a peace, a herd and a shepherd. An instrument of your body is also your little reason, my brother, which you call "spirit" — a little instrument and toy of your great reason.
>
> "I," you say, and are proud of the word. But greater is that in which you do not wish to have faith — your body and its great reason: that does not say "I," but does "I."
>
> What the sense feels, what the spirit knows, never has its end in itself. But sense and spirit would persuade you that they are the

end of all things: that is how vain they are. Instruments and toys are sense and spirit: behind them still lies the self. The self also seeks with the eyes of the senses; it also listens with the ears of the spirit. Always the self listens and seeks: it compares, over-powers, conquers, destroys. It controls, and it is in control of the ego too.

"Behind your thoughts and feelings, my brother, there stands a mighty ruler, an unknown sage — whose name is self. In your body he dwells; he is your body (Z, 34).

Here in one brief passage Nietzsche denies the existence of any separate "soul", affirms the existence only of the body, and interprets and praises the "great reason" which is this body. All future references to the human person are to be understood as this or that *body*. This is *who* you are! Note also in the passage ". . . and soul is only a word for something about the body." The small word "about" (considering Nietzsche's later emphasis on physiology and physiological psychology) is major. "About", meaning "around, on every side of," but also meaning "fundamentally concerned with or directed toward."

Earlier in his Preface for the Second Edition of *The Gay Science*, Nietzsche had written:

The unconscious disguise of physiological needs under the cloaks of the objective, ideal, purely spiritual goes to frightening lengths — and often I have asked myself whether, taking a large view, philosophy has not been merely an interpretation of the body and a *misunderstanding of the body*.

Behind the highest value judgments that have hitherto guided the history of thought, there are concealed misunderstandings of the physical constitution — of individuals or classes or even whole races. All those bold insanities of metaphysics, especially answers to the question about the *value* of existence, may always be considered first of all as the symptoms of certain bodies. And if such world affirmations or world negations *tout court* lack any grain of significance when measured scientifically, they are more valuable for the historian and psychologist as hints or symptoms of the body, of its success or failure, its plenitude, power, and autocracy in history, or of its frustration, weariness, impoverish-ment, its premonition of the end, its will to the end.

I am still waiting for a philosophical *physician* in the exceptional sense of that word — one who has to pursue the problem of the total health of a people, time, race or of humanity — to muster the courage to push my suspicion to its limits and to risk the proposition: what was at stake in all philosophizing hitherto was not at all "truth" but something else — let us say, health, future, growth, power, life (*GS*, 34, 35).

Why should we be surprised that in focusing on power, on life, on the body, Nietzsche also focused on sexuality, the sexual body, the male body, similar to and different from the female body? Remarkably, he was urging us to consider the hypothesis that deep down *everything* which is called psychological, or the psyche, is ultimately and always "about the body."

Later, in *Beyond Good and Evil*, Nietzsche suggested the following:

> Let us assume that nothing is "given" as real except our world of desires and passions, that we cannot step down or step up to any kind of "reality" except the reality of our drives — for thinking is nothing but the interrelation and interaction of our drives..
> .. Assuming finally, that we succeeded in explaining our entire instinctual life as the development and ramification of one basic form of will (of the will to power, as I hold); assuming that one could trace back all the organic functions of this will to power, including the solution of the problem of generation and nutrition (they are one problem) — if this were done, we should be justified in defining *all* effective energy unequivocally as *will to power*. The world seen from within, the world designated and defined according to its "intelligible character" — this world would be *will to power* and nothing else (*BGE*, 42, 43, 44).

Nietzsche's belief, his interpretation, that one single drive, which he called will to power, the drive for power, for the feeling of power, was the source and root of everything physiological and psychological, only increased and grew in strength until his untimely collapse. He made efforts to apply this hypothetical principle to multiple forms of human behavior and psychological phenomena, some simple, others complex. The evidence is abundant and persuasive that a few phenomena became more demanding, more persistent, more directly related to the drive for power. Three of these — fear, cruelty, and deceit — may have originally engaged Nietzsche's attention in his becoming familiar with the political philosophy of Machiavelli.

We are trudging our way to some possible different interpretation of war, and although considerable attention has been given thus far in our task to Machiavelli's thinking, it seems important to bring him again into the discussion. Machiavelli believed, if in a more constricted context, that the starting point of all philosophy was power. As a political philosopher, he understood that power and war were inextricably related. He wrote:

> A prince should therefore have no other aim or thought, nor take up any other thing for his study, but war and its organization and discipline, for that is the only art that is necessary to one who commands, and it is of such virtue that it not only maintains those who are born princes, but often enables men of private fortune to attain to that rank (*GPT*, 291).

While Machiavelli restricted his attention to power of the ruler, or the state, Nietzsche developed his own reconsidered ideas of fear, cruelty, and deceit and their relationship to power. Fear is best understood as a negative aspect of the drive to power, i.e. as the feeling of the lack of power. Also he wrote, "Here too fear is once again the mother of morality." And morality, remember, he defines as "the doctrine of rank-relations that produce the phenomenon we call life." Nietzsche also recognized that fear of the unknown drives toward the acquisition of knowledge. He suggests that fear of the senses was typical of earlier philosophers, fear that evidence of the senses could challenge or contradict their "rational" ideas. He said that having once learned to fear men, women now had "unlearned" this fear. There is the suggestion that perhaps fear is the most active of the passions, and always is interdependently related to power. Nietzsche considers other passions, but fear is fundamental in addressing the drive for power.

Nietzsche's sensibility to the cruelty in the world was, as with many other aspects, exceptional. Here is one example:

> There is a great ladder of religious cruelty. It has many rungs, but three of them are of the greatest importance. The first is the sacrifice of men to one's God, perhaps those men in particular whom one most loved. Among these are the first-born offerings of all primitive religions, and also Emperor Tiberius' sacrifice in the Mithras grotto on the Isle of Capri — that most gruesome of all Roman anachronisms. The second rung, attained in the moral period of mankind, is the sacrifice to one's god of one's strongest instincts, one's "natural man." That sort of festive joy gleams in the cruel eyes of the ascetics, the enthusiastic "anti-naturalists." And finally — what remains that could be sacrificed? Don't we in the end have to sacrifice everything consolatory, holy, and healing; all hope, all belief in invisible harmony, in future blessedness and justice? Don't we have to sacrifice God himself and idolize a rock, the forces of stupidity, of gravity, fate, nothingness — all in order to be sufficiently cruel to ourselves? To sacrifice God for

nothingness — this is the paradoxical mystery of ultimate cruelty that remained in store for the generation now growing up. All of us know something about it already (*BGE*, 61, 62).

In Nietzsche's view, as with Machiavelli's, cruelty is one major expression of the drive for power, an instrument of acquiring, maintaining, and increasing power. In the agon which is life, cruelty — mental and physical — mixed with fear, is formidable. *Fear* — the feeling, or awareness, of the lack of power. *Cruelty* — the inflicting of suffering, injury, grief, pain as an expression of power, or instrument for acquiring power. The significance of *deceit* will come later.

When Nietzsche pursued further his "deep psychology", his venture into the pre-conscious, or perhaps the unconscious, he discovered something more concealed, more astonishing and alarming, than the more obvious phenomena, fear and cruelty. He discovered *ressentiment*. In my recent book, *Nietzsche, Philosopher of the Perilous Perhaps*, there is a detailed discussion of the nature and significance of the psychology of ressentiment for Nietzsche, especially in developing his critique of Christianity. We need to revisit Nietzsche's own words. In his *Genealogy of Morals* he writes:

> All men of resentment are these physiologically distorted and worm-riddled persons, a whole quivering kingdom of burrowing revenge, indefatigable and insatiable in its outburst against the happy, and equally so in disguises for revenge, in pretexts for revenge: when will they really reach their final, fondest, most sublime triumph of revenge? At that time, doubtless, when they succeed in pushing their own misery, indeed all misery there is, into the *consciousness* of the happy; so that the latter begin one day to be ashamed of their happiness, and perchance say to themselves when they meet, "It is a shame to be happy! *There is too much misery!*" (E, 179)

In one of his last books, *The Antichrist*, we read this:

> In my *Genealogy of Morals* I offered the first psychological analysis of the counter concepts of a *noble* morality and a morality of *ressentiment* — the latter born of the No to the former: but this is the Judaeo-Christian morality, pure and simple. So that it could say No to everything on earth that represents the ascending tendency of life, to that which has turned out well, to power, to beauty, to self-affirmation, the instinct of *ressentiment*, which had here become genius, had to invent *another* world from whose point of view this affirmation of life appeared as evil, as the reprehensible as such . . . (*PN*, 593).

Finally, we have some of Nietzsche's last deliberations in *Ecce Homo*:

> Freedom from *ressentiment*, enlightenment over *ressentiment* — who knows the extent to which I ultimately owe thanks to my pro-tracted sickness for this too! The problem is not exactly simple: one has to have experienced it from a state of strength and a state of weakness. If anything whatever has to be admitted against be-ing sick, being weak, it is that in these conditions the actual cura-tive instinct, that is to say the *defensive* and *offensive instinct* in man becomes soft. One does not know how to get free of anything, one does not know how to have done with anything, one does not know how to thrust back — everything hurts. Men and things come importunately close, events strike too deep, the memory is a festering wound. Being sick *is* itself a kind of *ressentiment* . . . And nothing burns one up quicker than the affects of *ressentiment*. Vexation, morbid susceptibility, incapacity for revenge, poison-brewing in any sense — for one who is exhausted this is certainly the most disadvantageous kind of reaction: it causes a rapid ex-penditure of nervous energy, a morbid accretion of excretions, for example, of gall into the stomach. *Ressentiment* is the forbidden *in itself*, for the invalid — *his* evil: unfortunately also his most natural inclination. — This was grasped by that profound physiologist Buddha. His "religion", which one would do better to call a *system of hygiene* so as not to mix it up with such pitiable things as Chris-tianity, makes its effect dependent on victory over *ressentiment*: to free the soul of *that* — first step to recovery. "Not by enmity is enmity ended, by friendship is enmity ended": this stands at the beginning of Buddha's teaching — it is *not* morality that speaks thus, it is physiology that speaks thus. — *Ressentiment*. born of weakness, to no one more harmful than to the weak man himself. . . He who knows the seriousness with which my philosophy has taken up the struggle against the feelings of revengefulness and vindictiveness . . . my struggle against Christianity is only a spe-cial instance of it (*EH*, 45, 46).

Nietzsche clearly and repeatedly insisted that the source of *ressenti-ment*, and then of revenge, is the inability or unwillingness to accept oneself, to accept one's "fate," to affirm one's place in nature. From *Thus Spoke Zarathustra*, this was his plea:

> For *that man be delivered from revenge*, that is for me the bridge to the highest hope, and a rainbow after long storms (*Z*, 99).

A few further comments which apply to the psychological phe-nomenon of resentment as obviously it increased in significance for Nietzsche. Resentment appears to be deeply rooted historically in man's biological and existential situation. Resentment is directed primordially, however, toward the essential nature, the natural power, the physiology of the female in the process of producing new life, and the value of that natural, powerful ability. Although there are many situations or expressions of resentment, the archetypal expression is sexual — biological, physiological. Resentment, as the source, led to the emergence and development of what Nietzsche calls "slave-morality", i.e. Christian morality, which meant in Nietzsche's interpretation, to a distortion of a *natural order of values*. What had occurred in history was a reversal of values, a process of devaluing of the *woman* and the *child* — resulting, in Nietzsche's view, in a *false table of values*.

Nietzsche saw that resentment is a form of "self-poisoning." Perhaps deeply repressed, it leads to revenge, to violence. As it grows, it tends to have a distorting influence on perception, on thinking. And constituting Nietzsche's distress was his growing awareness that he was investigating, and coming to further understand, not only power, values, sexuality, fear, cruelty, but also resentment, revenge, and deception. Consider again — resentment as meaning "a feeling of indignant displeasure or persistent ill-will at something regarded as an insult, a wrong, an injury"; or, "an emotional response to a perceived indignity or insult, usually long-lasting"; or, "smoldering indignation." And revenge as meaning "inflicting injury in retaliation for a perceived insult, often by extreme or intense force, often sudden, violent, destructive, causing injury or death." Perhaps Nietzsche believed that humanity has more to fear from resentment than from fear itself.

And what about *deception*? If possible, perhaps this idea equaled the idea of power in the development of Nietzsche's thinking. Surely that was the case in his effort to expose the destructive nature and exercise of power in Christianity. And we may anticipate a similar significance in addressing the destructive nature and exercise of power in *war*.

But first, take a brief excursion into the language of deception, and the associated richness of meanings. Deception usually is understood as the art or practice of deceiving, concealing, or distorting of the truth for the purpose of misleading; duplicity, fraud, cheating, trickery. Familiar related terms include fraud, meaning deceit perpetrated to gain dishonest advantage; corruption, meaning dishonest practice, as bribery; crooked, untrustworthy; hypocrisy, meaning pretending to be what one is not. Other associations include dissimulation, artifice, stratagem, venality. On the opposite side — honesty, justice, trustworthiness, integrity, probity, veracity, rectitude, fairness. And two of Nietzsche's special terms — fabricate, or fabrication, meaning to make up for the purpose of deception. Also, mendacity or mendacious, meaning fabricating, inventing or telling lies, especially habitually; also meaning untruthfulness, dishonesty. Falsehood usually is considered as an untrue idea, or belief. Untruth suggests an incorrect statement that distorts or depresses, either intentionally or misleading — a lie. A lie is usually taken to mean a false statement made with deliberate intent to deceive, a vicious falsehood. Untruth, often used by Nietzsche, is less harsh than either falsehood or lie, and may arise from misunderstanding or ignorance. The idea, or word, untruth, also has much broader applications or uses — fable, fiction, story, legend, myths, poetry, symbols, and more.

In Nietzsche's interpretations, the subjectivity and relativity of *truth* was a starting point, a given. What is called "truth" is always shifting, changing, never absolute or eternal. Probably the notion of what is to be called "untruth" is similarly fluid. A sampling of Nietzsche's own words may provide a suggestion of the extent and depth of his perspectives and reflections. From his early *Human, All Too Human*, these:

> *The point of honesty in deception.* In all great deceivers there occurs a noteworthy process to which they owe their power. In the actual act of deception, among all the preparations, the horror in the voice, expression, gestures, amid the striking scenery, the *belief in themselves* overcomes them. It is this that speaks so miraculously and convincingly to the onlookers. The founders of religions are distinguished from those other great deceivers by the fact that they do not come out of this condition of self-deception: or, very infrequently, they do have those clearer moments, when doubt overwhelms them, but they usually comfort themselves by foisting these clearer moments off on the evil adversary. Self-deception must be present, so that both kinds of deceivers can have a grand *effect*. For men will believe something is true, if it is evident that others believe in it firmly. (*HA*, 51)

> *The lie.* Why do men usually tell the truth in daily life? Certainly not because a god has forbidden lying. Rather it is because, first, it is more convenient; for lies demand imagination, dissembling, and memory (which is why Swift says that the man who tells a lie seldom perceives the heavy burden he is assuming: namely, he must invent twenty other lies to make good the first). Then, it is

because it is advantageous in ordinary circumstances to say directly: I want this, I did that, and so on; that is, because the path of obligation and authority is safer than that of cunning.

If a child has been raised in complicated domestic circumstances, however, he will employ the lie naturally, and will always say instinctively that which corresponds to his interests. A feeling for truth, a distaste for lying in and out of itself, is alien to him and inaccessible; and so he lies in complete innocence (*HA*, 52).

Enemies of truth. Convictions are more dangerous enemies of truth than lies (*HA*, 234).

Truth. No one dies of fatal truth nowadays: there are too many anecdotes (*HA*, 238).

Conviction is the belief that in some point of knowledge one possesses absolute truth. Such a belief presumes, then, that absolute truths exist; likewise, that the perfect methods for arriving at them have been found; finally, that every man who has convictions makes use of these perfect methods. All three assertions prove at once that the man of convictions is not the man of scientific thinking; he stands before us still in the age of theoretical innocence, a child, however grownup he might be otherwise. But throughout thousands of years, people have lived in such childlike assumptions, and from out of them mankind's mightiest sources of power have flowed. The countless people who sacrificed themselves for their convictions thought they were doing it for absolute truth. All of them were wrong; probably no man has ever sacrificed himself for truth; at least, the dogmatic expression of his belief will have been unscientific, half-scientific.... It is not the struggle of opinions that has made history so violent, but rather the struggle of belief in opinions, that is, the struggle of convictions. If only all those people who thought so highly of their conviction, who sacrificed all sorts of things to it and spared neither their honor, body nor life in its service, had devoted only half of their strength to investigating by what right they clung to this or that conviction, how they had arrived at it, then how peaceable the history of mankind would appear! How much more would be known! All the cruel scenes during the persecution of every kind of heretic would have been spared us for two reasons: first, because the inquisitors would above all have inquired within themselves, and got beyond the arrogant idea that they were defending the absolute truth; and second, because the heretics themselves would not have granted such poorly established tenets as those of all the sectarians and "orthodox" any further attention, once they had investigated them (*HA*, 261, 262).

Later, from *The Gay Science*, here are a few excerpts:

Those who feel "I possess Truth" — how many possessions would they not abandon in order to save this feeling! What would they not throw overboard to stay "on top" — which means *above* the others who lack "the Truth"! (*GS*, 87)

Not predestined for knowledge. — There is stupid humility that is not at all rare, and those afflicted with it are altogether unfit to become devotees of knowledge. As soon as a person of this type perceives something striking, he turns on his heel, as it were, and says to himself: "You have made a mistake. What is the matter with your senses? This cannot, may not, be the truth." And then, instead of looking and listening again, more carefully, he runs away from the striking thing, as if he had been intimidated, and tries to remove it from his mind as fast as he can. For his inner canon says: "I do not want to see anything that contradicts the prevalent opinion. Am *I* called to discover new truths? There are too many old ones, as it is" (*GS*, 100).

Origin of knowledge. — Over immense periods of time the intellect produced nothing but errors. A few of these proved to be useful and helped to preserve the species: those who hit upon or inherited these had better luck in their struggle for themselves and their progeny. Such erroneous articles of faith, which were continually inherited, until they became almost part of the basic endowment of the species, include the following: that there are enduring things; that there are equal things; that there are things, substances, bodies; that a thing is what it appears to be; that our will is free; that what is good for me is also good in itself. It was only very late that such propositions were denied and doubted; it was only very late that truth emerged — as the weakest form of knowledge. It seemed that one was unable to live with it: our organism was prepared for the opposite; all its higher functions, sense perception and every kind of sensation worked with those basic errors, which had been incorporated since time immemorial. Indeed, even in the realm of knowledge these propositions became the norms according to which "true" and "untrue" were determined — down to the most remote regions of logic. . . . Gradually the human brain became full of such judgments and convictions, and a ferment, struggle, and lust for power developed in this tangle. Not only utility and delight but every kind of impulse took sides in this fight about "truths." The intellectual fight became an occupation, an attraction, a profession, a duty, something dignified — and eventually knowledge and the striving for the true found their place as a need among other needs. Henceforth not only faith and conviction but also scrutiny, denial, mistrust, and contradiction became a *power* . . . . Thus knowledge became a piece of life itself, and hence a continually growing power — until eventually knowledge collided with those primeval errors: two lives, two powers, both in the same human being. A thinker is now that being in whom the impulse for truth and those life-preserving errors clash for their first fight, after the impulse for truth has proved to be also a life-preserving power. Compared to the significance of this fight, everything else is a matter of indifference: the ultimate question about the conditions of life has been posed here, and we confront the first attempt to answer this question by experiment. To what extent can truth endure incorporation? That is the question; that is the experiment (*GS*, 169, 170, 171).

Multiplication table. — One is always wrong, but with two, truth begins. — *One* cannot prove his case, but two are irrefutable (*GS*, 218).

Ultimate skepsis. — What are man's truths ultimately? Merely his *irrefutable* errors (*GS*, 219).

In favor of criticism. — Now something that you formerly loved as a truth or probability strikes you as an error; you shed it and fancy that this represents a victory for your reason. But perhaps this error was as necessary for you then, when you were still a different person — you are always a different person — as are all your present "truths," being a skin, as it were, that concealed and covered a great deal that you were not yet permitted to see. What killed that opinion for you was your new life and not your reason: *you no longer need it,* and now it collapses and unreason crawls out of it into the light like a worm. When we criticize something, this is no arbitrary and impersonal event; it is, at least very often, evidence of vital energies in us that are growing and shedding a skin. We negate and must negate because something in us wants to live and affirm — something that we perhaps do not know or see as yet. — That is said in favor of criticism (*GS*, 245, 246).

As interpreters of our experiences. — One sort of honesty has been alien to all founders of religion and their kind: They have never made their experiences a matter of conscience for knowledge. "What did I really experience? What happened in me and around me at that time? Was my reason bright enough? Was my will opposed to all deceptions of the senses and bold in resisting the fantastic?" None of them has asked such questions, nor do any of our dear religious people ask them even now. On the contrary, they thirst after things that *go against reason*, and they do not wish to make it too hard for themselves to satisfy it. So they experience "miracles" and "rebirths" and hear the voices of little angels! But we, we others who thirst after reason, are determined to scrutinize our experiences as severely as a scientific experiment — hour after hour, day after day. We ourselves wish to be our experiments and guinea pigs (*GS*, 253).

The question whether *truth* is needed must not only have been affirmed in advance, but affirmed to such a degree that the principle, the faith, the conviction finds expression. *Nothing* is needed *more* than truth, and in relation to it everything else has only second-rate value.

This unconditional will to truth — what is it? Is it the will *not to allow oneself to be deceived*? Or is it the will *not to deceive*? For the will to truth could be interpreted in the second way, too — if only the special case "I do not want to deceive myself" is subsumed under the generalization "I do not want to deceive." But why not deceive? But why not allow oneself to be deceived?

Note that the reasons for the former principle belong to an altogether different realm from those for the second. One does not want to allow oneself to be deceived because one assumes that it is harmful, dangerous, calamitous to be deceived. In this sense, science would be a long-range prudence, a caution, a utility; but one could object in all fairness: How is that? Is wanting not to allow oneself to be deceived really less harmful, less dangerous, less calamitous? What do you know in advance of the character of existence to be able to decide whether the greater advantage is on the side of the unconditionally mistrustful or of the unconditionally trusting? But if both should be required, much trust *as well as* much mistrust, from where would science be permitted its unconditional faith or conviction on which it rests, that truth is more important than any other thing, including every other conviction? Precisely this conviction could never have come into being if both truth and untruth constantly proved to be useful, which is the case. . . .

Consequently, "will to truth" does *not* mean "I will not allow myself to be deceived" but — there is no alternative — "I will not deceive, not even myself"; *and with that we stand on moral ground (GS, 281, 282).*

By the time Nietzsche was entering the explosive period of the last couple of years of his creative energies, even more of his attention was being drawn to the burgeoning demands of the issue of truth. His Preface to *Beyond Good and Evil* begins:

Supposing that Truth is a woman — well, now, is there some foundation for suspecting that all philosophers, insofar as they were dogmatists, have not known how to handle women? That the gruesome earnestness, the left-handed obtrusiveness, with which they have usually approached Truth have been unskilled and unseemly methods for prejudicing a woman (of all people!) in their favor? One thing is certain: she has not been so prejudiced. Today, every sort of dogmatism occupies a dismayed and discouraged position — if, indeed, it has maintained any position at all. For there are scoffers who maintain that dogma has collapsed, even worse, that it is laboring to draw its last breath. Seriously speaking, there are good grounds for hoping that all philosophic dogmatizing, however solemn, however final and ultimate it has pretended to be, may after all have been merely a noble child's-play and mere beginning. And perhaps the time is very near when we shall again and again comprehend *how* flimsy the cornerstone has been upon which the dogmatists have hitherto built their sublime and absolute philosophical edifices (*BGE*, xi).

The First Article in the same book begins:

The will to truth! That will which is yet to seduce us into many a venture, that famous truthfulness of which all philosophers up to this time have spoken reverently — think what questions this will to truth has posed for us! What strange, wicked, question-able questions! It has been a long story — and yet it seems hardly to have started. No wonder if just for once we become suspicious, and, losing our patience, impatiently turn around! Let us learn to ask this Sphinx some questions ourselves, for a change. Just *who* is it anyway who has been asking these questions? Just *what* is it in us that wants "to approach truth"? Indeed, we tarried a long time before the question of the cause of this will. And in the end we stopped altogether before the even more basic question. We asked "What is the value of this will?" Supposing we want truth: *why not rather* untruth? Uncertainty? Even Ignorance? The prob-lem of the value of truth confronted us — or were we the ones who confronted the problem?...

How is it possible for anything to come out of its opposite? Truth, for example, out of error? Or the will to truth out of the will to deception? Or a selfless act out of self-interest? Or the pure sunny contemplation of a wise man out of covetousness? This sort of or-igin is impossible. Who dreams of it is a fool or worse; the things of highest value must have some other, *indigenous* origin; they can-not be derived from this ephemeral, seductive, deceptive, inferior world, this labyrinth of delusion and greed! (*BGE*, 1, 2)

A few paragraphs later:

To admit untruth as a necessary condition of life: this implies, to be sure, a perilous resistance against customary value-feel-ings. A philosophy that risks it nonetheless, if it did nothing else, would by this alone have taken its stand beyond good and evil (*BGE*, 4).

Here are few notes from *The Will To Power*:

Critique of the holy lie. — That the lie is permitted as a means to pious ends is part of the theory of every priesthood — to what extent it is part of their practice is the object of this inquiry.

But philosophers, too, as soon as, with priestly ulterior motives, they form the intention of taking in hand the direction of man-kind, at once also arrogate to themselves the right to tell lies: Plato before all.. . .

The holy lie therefore invented (1) a *God* who punishes and re-wards, who strictly observes the law-book of the priests and is strict about sending them into the world as his mouthpieces and plenipotentiaries; (2) an *afterlife* in which the great punishment machine is first thought to become effective — to this end the *immortality of the soul*; (3) *conscience* in man as the consciousness that good and evil are permanent — that God himself speaks through it when it advises conformity with priestly precepts; (4) *morality* as denial of all natural processes, as reduction of all events to a morally conditioned event, moral effects (i.e., the idea of punishment and reward) as effects permeating all things, as the sole power, as the creator of all transformation; (5) *truth* as given, as revealed, as identical with the teaching of the priests: as the condition for all salvation and happiness in this life and the next (*WP*, 89, 90, 91).

We find a species of man, the priestly, which feels itself to be the norm, the high point and the supreme expression of the type man: this species derives the concept "improvement" from itself. It believes in its own superiority, it wills itself to be superior in fact: the origin of the holy lie is the *will to power*.

Establishment of rule: to this end, the rule of those concepts that place a *non plus ultra* of power with the priesthood. Power through the lie — in the knowledge that one does not possess it , physi-cally, militarily — the lie as a supplement to power, a new con-cept of "truth."

It is a mistake to suppose an *unconscious and naïve* development here, a kind of self-deception — Fanatics do not invent such carefully thought-out systems of oppression — The most cold-blooded reflection was at work here; the same kind of reflection as a Plato applied when he imagined his "Republic." "He who wills the end must will the means" — all lawgivers have been clear in their minds regarding this politician's insight (*WP*, 92).

The great lie in history: as if it was the *corruption* of paganism which opened the road to Christianity! It was, on the contrary, the weakening and *moralization* of the man of antiquity! Natural drives had already been reinterpreted as vices! (*WP*, 94)

Great lie in history: as if the corruption of the church were the cause of the Reformation! Only pretext and self-deception on the part of the instigators — strong needs were present whose bru-tality very much required a spiritual cloak (*WP*, 205).

These are only a few of Nietzsche's thoughts on truth, and espe-cially on untruth. It was evident that truth, or the lack of truth, was inextricably related to power, to values, to sexuality, to fear, to cruelty, and to resentment — and more.

In the Foreword to *Ecce Homo* one of his final conclusions in his long struggle against Christianity reads:

> Reality has been deprived of its value, its meaning, its veracity to the same degree as an ideal world has been *fabricated* . . . The "real world" and the "apparent world" — in plain terms: the *fabricated* world and reality. . . The *lie* of the ideal has hitherto been the curse on reality, through it mankind itself has become mendacious and false down to its deepest instincts — to the point of worshipping the *inverse* values to those which alone could guarantee it prosperity, future, the exalted *right* to a future. (*EH*, 34)

PART FIVE

The preliminaries have been prepared and the background has been sketched. Now for the main encounter. From early beginnings Nietzsche set as his target the religion of Christianity, and as his task (which occupied almost all of his entire creative life) that of "the unmasking of Christian morality." In his Preface to *Beyond Good and Evil*, he wrote:

> (The Germans invented gunpowder — all credit to them! But they made up for it by inventing the printing press.) We, however, who are neither Jesuits nor democrats nor even very German, we good *Europeans* and free, *very* free thinkers — we have it still, all the necessitation of spirit and all the tension of the bow! And perhaps also the arrow, the task, and (who knows?) the *target*. . . .(*BGE*, xiii).

Nietzsche's friend, Peter Gast, wrote to him:

> [Y]ou have driven your artillery on the highest mountain, you have such guns as have never yet existed, and you need only shoot blindly to inspire terror all around. . . (*PN*, 464).

Nietzsche's *war*, or one of them, was a war against the values invented and nurtured within the doctrines of Christianity in Western culture. He wrote this:

> *Who is most influential* — .When a human being resists his whole age and stops it at the gate to demand an accounting, this *must*

have influence. Whether that is what he desires is immaterial; that he can do it is what matters (*GS*, 198).

The Christian narrative, the ideology, was in his view not merely an invented story, a fiction, it was a *lie*, intended to mislead and deceive.

From Nietzsche's many perspectives, the sense of war — of strife, competition, conflict, struggle, between opposing forces — was constantly occurring in many situations. It seems important to identify the major ones. In *Human, All Too Human*, he wrote this:

> *Pleasure in knowing.* Why is knowledge, the element of researchers and philosophers, linked to pleasure? First and foremost, because by it we gain awareness of our power — the same reason that gymnastic exercises are pleasurable even without spectators. Second, because, as we gain knowledge, we surpass older ideas and their representatives, become victors, or at least believe ourselves to be. Third, because any new knowledge, however small, makes us feel superior to *everyone* and unique in understanding this matter correctly. These three reasons for pleasure are the most important, but depending on the nature of the knower, there are still many secondary reasons (*HA*, 154).

Commenting on this passage, Hollingdale reminded us:

> The idea behind this passage is of course that, to put it at its broadest, life is a contest, and that the agonistic character of human existence extends into fields where one would not necessarily expect to find it. That the *agon* was the model for all human activity had already suggested itself to Nietzsche during his studies of Greek culture and civilization, where it seemed to him to stand out clearly as a fact. His conclusions had been summarized in "Homer's Contest", the fifth of the *Five Prefaces to Five Unwritten Books* (1872), where he drew attention to the opening of Hesiod's *Works and Days* — the passage in which Hesiod says there are *two* Eris-goddesses on earth, one the promoter of war, the other "much better one" the promoter of competition — and commented: "The whole of Greek antiquity thinks of spite and envy otherwise than we do, and judges as does Hesiod, who first designates as evil that Eris who leads men against one another to a war of extermination, and then praises another Eris as good who, as jealousy, spite, envy, rouses men to deeds, but not to deeds of war but to deeds of *contest*." The conception thus summarized in an unpublished manuscript achieved publication in *Human, All Too Human* in the form: "The Greek artists, the tragedians for example, poetized in order to conquer; their entire art is inconceivable without contest: Hesiod's good Eris, ambition, gave their genius wings". . . This agonistic interpretation of culture is extended, naturally enough, to include the realm of philosophy, and it is done so in a very whole-hearted and comprehensive way:

the Greek philosophers, Nietzsche writes, "had a sturdy faith in themselves and their 'truth' and with it they overthrew all their contemporaries and predecessors; every one of them was a violent and warlike *tyrant*" (N, 78, 79).

Further into his discussion, Hollingdale says:

> A state of conflict, external and internal, is the pre-condition for "growth", "enhancement", in short for *Ubermensch*: this dictum is reiterated in a hundred different ways and at all periods of Nietzsche's life. "Strife is the perpetual food of the soul", he wrote in 1862, at the age of seventeen; "One must still have chaos in one to be able to give birth to a dancing star", said Zarathustra twenty-one years later. . . In *Beyond Good and Evil*, internal "war" is, together with the capacity for its direction, posited as the force which creates genius:. . (N, 109).

Hollingdale then quotes from his own translation of *Beyond Good and Evil* as follows:

> The man of an era of dissolution which mixes the races together and who therefore contains within him the inheritance of a diversified descent, that is to say contrary and often not merely contrary drives and values which struggle with one another and rarely leave one another in peace — such a man. . . will, on average, be a rather weak man: his fundamental desire is that the war which he *is* should come to an end. . .If, however, the contrariety and war in such a nature should act as one *more* stimulus and enticement to life — and if, on the other hand, in addition to powerful and irreconcilable drives, there has also been inherited and cultivated a proper mastery and subtlety in conducting a war against oneself, that is to say self-control, self-outwitting: then there arise those marvelously incomprehensible and unfathomable men, those enigmatic men predestined for victory and the seduction of others, the fairest examples of which are Alcibiades and Caesar. . . and among artists perhaps Leonardo da Vinci (N, 109).

External and internal conflict defined Nietzsche's life. He was a born "warrior." It was his fate. Instincts, beliefs, ideas, passions, desires, values — often contrary, always challenging, always struggling. Desiring to be heard and understood, as well as misunderstood. He recognized, appreciated, and valued many forms of competition very different from overt acts of violence. Hesiod's "good Eris" was his goddess.

Some of Nietzsche's last deliberations brought this in *Ecce Homo*:

> War is another thing. I am by nature warlike. To attack is among my instincts. *To be able* to be an enemy, to be an enemy — that perhaps presupposes a strong nature, it is in any event a condi-

tion of every strong nature. It needs resistances, consequently it *seeks* resistances: . . for a philosopher who is warlike also challenges problems to a duel. The undertaking is to master, *not* any resistances that happen to present themselves, but those against which one has to bring all one's strength, suppleness and mastery of weapons (*EH*, 47).

Earlier in his career Nietzsche had been pointing ahead to later bolder claims. In *The Gay Science*, he suggested this:

Good luck in fate. — The greatest distinction that fate can bestow on us is to let us fight for a time on the side of our opponents. With that we are *predestined* for a great victory (*GS*, 255).

And this typical passage:

In media vita. — No, life has not disappointed me. On the contrary, I find it truer, more desirable and mysterious every year — ever since the day when the great liberator came to me: the idea that life could be an experiment of the seeker for knowledge — and not a duty, not a calamity, not trickery. — And knowledge itself: let it be something else for others; for example, a bed to rest on, or the way to such a bed, or a diversion, or a form of leisure — for me it is a world of dangers and victories in which heroic feelings, too, find places to dance and play. *"Life as a means to knowledge"* — with this principle in one's heart one can live not only boldly but even gaily, and laugh gaily, too. And who knows how to laugh anyway and live well if he does not first know a good deal about war and victory? (*GS*, 255)

Back to *Ecce Homo*. Nietzsche finally gives voice to another "sense" of war, one which had been evolving and gaining attention almost from the very earliest stages of his experiences. These carefully crafted words:

The more a woman is a woman the more she defends herself tooth and nail against rights in general: for the state of nature, the eternal *war* between the sexes puts her in a superior position by far (*EH*, 76).

Among the opening words of the last chapter of *Ecce Homo* are these:

I KNOW my fate. One day there will be associated with my name the recollection of something frightful — of a crisis like no other before on earth, of the profoundest collision of conscience, of a decision evoked *against* everything that until then had been believed in, demanded, sanctified. . . . I was the first to sense — *smell* — the lie as lie. . . My genius is in my nostrils. . . I contradict as has never been contradicted and am nonetheless the opposite of a negative spirit. I am a *bringer of good tidings* such as there has never been, I know tasks from such a height that any conception of them has hitherto been lacking; only after me is it possible to hope again. With all that I am necessarily a man of fatality. For when truth steps into battle with the lie of millennia we shall have convulsions, an earthquake spasm, a transposition of valley and mountain such as has never been dreamed of. The concept politics has then become completely absorbed into a war of spirits, all the power-structures of the old society have been blown into the air — they one and all reposed on the lie; there will be wars such as there have never yet been on earth. Only after me will there be *grand politics* on earth. (EH, 126, 127)

Nietzsche was fully engaged his entire life in war, as a warrior — in conflict, in competition. War, or wars, meant essentially the war of ideas, of words, regarding *values* and *truth* or *lies*. He spoke of these throughout his turbulent experiences. War, in his interpretations, had a related second meaning. Dealt with in his writings more surreptitiously, indirectly, and cleverly was the sense of war between the sexes — conflict between the value of the powers, i.e., abilities of the male versus those of the female. In Nietzsche's world, both the war between ideas and the war between the sexes were *eternal*. Remember, life — the process of living life — is agonistic, always, everywhere.

And war, the killing and dying kind of war, the "real" war? We know Nietzsche used the strongest words to portray his perspectives on German culture and on Western culture. Decaying, decadent, deteriorating, declining. In one of a series of university lectures in 1951 and 1952 on Nietzsche's thinking, reproduced in his book, *What Is Called Thinking*, Martin Heidegger says this:

A generation ago it was "The Decline of the West." Today we speak of "loss of center." People everywhere trace and record the decay, the destruction, the imminent annihilation of the world. We are surrounded by a special breed of reportorial novels that do nothing but wallow in such deterioration and depression. On the one hand, that sort of literature is much easier to produce than to say something that is essential and truly thought out; but on the other hand it is already getting tiresome. The world, men find, is not just out of joint but tumbling away into the nothingness of absurdity. Nietzsche, who from his supreme peak saw far ahead of it all, as early as the eighteen-eighties had for it the simple, because thoughtful words: "The wasteland grows." It means, the devastation is growing wider. Devastation is more than destruction. Devastation is more unearthly than destruction. Destruction only sweeps aside all that has grown up or been built up so far; but devastation blocks all future growth and prevents all building. Devastation is more unearthly than destruction (*CT*, 29).

In another of these lectures, Heidegger repeats Nietzsche's words in this important way, "The wasteland grows; woe to him who hides wastelands within!" Heidegger reminds us also of the important fact that Nietzsche's demands that he himself, and we, in thinking and re-thinking, must commit to a return to *origins*, *sources*, *roots* — to seek a new beginning for understanding. This approach, or method, took Nietzsche directly to the question of the source of this cultural devas-tation. And his answer was unequivocally, to religion — specifically Christianity. There is reliable evidence that Nietzsche became con-vinced, as was Bachofen, that the primary force shaping any culture is religion, past and present. Nietzsche extended that to mean that the powers of the religion of Christianity had resided in its successful invention and propagation of a system or "table of values" — values which Nietzsche interpreted as negative, reversals, devaluations of ear-lier original values — of nature, of life, the body, sexualilty, the woman — ancient religious values. And Christian values were beginning to be recognized as failed, false, destructive. The deteriorating condition of Western culture was the result of a strategy, an elaborate narrative, a lie, the aim of which was to permanently alter the power and value rela-tions between females and males — the prototypical pair of contraries in nature — the basis of that phenomenon we call "life."

Nietzsche conceived of his immediate and necessary task as devot-ing all of his own powers to exposing the true nature of Christianity, the lies, and the past, present, and possible future consequences of these lies. But a deeper concern with origins, or sources, became evi-dent in the question of the source of Christianity and the ideas and values which were at the center of this Christian narrative. Recogniz-ing this as a question for the psychologist, Nietzsche's answer was again unequivocal. From many perspectives — experiential, histori-cal, physiological, philosophical, experimental — his "discovery" was *resentment*, accompanied or expressed by *revenge*. And, the inventing and implementing of the Christian narrative — in Nietzsche's view, the lie or lies — was the attempt to disguise the underlying resentment, to achieve revenge. Nietzsche was driven inexorably and reluctantly to scream and whisper, to warn of the danger to human life, of this "poison." Resentment and revenge continue to recur in thinking our way to confronting war, as they did in Nietzsche's thinking. In one of the speeches of Nietzsche's Zarathustra, he spoke:

> Behold, this is the hole of the tarantula. Do you want to see the tarantula itself? Here hangs its web; touch it, that it tremble!
>
> There it comes willingly: welcome, tarantula! Your triangle and symbol sits black on your back; and I also know what sits in your soul. Revenge sits in your soul: wherever you bite, black scabs grow; your poison makes the soul whirl with revenge.
>
> Thus I speak to you in a parable — you who make souls whirl, you preachers of *equality*. To me you are tarantulas, and secretly vengeful. But I shall bring your secrets to light; therefore I laugh in your faces with my laughter of the heights. Therefore I tear at your webs, that your rage may lure you out from behind your word justice. For, *that man be delivered from revenge*, that is for me the bridge to the highest hope, and a rainbow after long storms.
> . . .
>
> Although they are sitting in their holes, these poisonous spiders, with their backs turned on life, they speak in favor of life, but only because they wish to hurt. They wish to hurt those who now have power, for among these the preaching of death is still most at home (*Z*, 99, 101).

In another speech, Zarathustra reminds again:

> *The spirit of revenge*, my friends, has so far been the subject of man's best reflection; and where there was suffering, one always wanted punishment too. "For 'punishment' is what revenge calls itself; with a hypocritical lie it creates a good conscience for itself (*Z*, 140).

In Part One of this book, I quoted from William Ebenstein's book, *Great Political Thinkers* as follows:

> In 1932 the League of Nations and the International Institute of Intellectual Cooperation asked Albert Einstein, the greatest physicist since Newton, to choose a significant issue of the time, and to discuss it with a person to be selected by him. Einstein chose the problem: "Is there a way of delivering mankind from the menace of war?" as the most important issue of humanity, and he selected Freud as the person most qualified to illuminate its cause and cure. This exchange of letters was published by the League of Nations under the title of *Why War?* in 1933. By that time the establishment of the Nazi regime in Germany had made the question one of tragic timeliness (*GPT*, 837).

Einstein lived to experience the second great war in Europe, and beyond. He believed, as others before and after him, that ultimately understanding war would lead us to psychologists. And Einstein chose to ask the advice of Sigmund Freud. As the most important issue of his time, Einstein wrote to Freud:

> This is the problem: Is there any way of delivering mankind from the menace of war? It is common knowledge that, with the advance of modern science, the issue has come to mean a matter of life and death for civilisation as we know it; nevertheless, for all the zeal displayed, every attempt at its solution has ended in a lamentable breakdown (GPT, 851).

Einstein understandably identified the matter of "life and death" as "delivering mankind from the menace of war." Freud from his limited psychological perspectives, was unable to offer Einstein any solace, or any solution. Supposing Einstein had selected not Freud, but Nietzsche, or the "spirit" of Nietzsche, for answers to "this urgent and absorbing question." A few decades earlier, in Germany, Nietzsche had identified the greatest issue facing humanity as "that man be delivered from revenge." And that meant also from resentment, which is the source of revenge. In order to think in new ways, one must have achieved freedom from the ravages of resentment and revenge. Only then may one become what Nietzsche calls a "free-thinker." Heidegger spoke as follows:

> Nietzsche's thinking focuses on deliverance from the spirit of revenge. It focuses on a spirit which, being the freedom from revenge, is prior to all mere fraternization, but also to any mere desire to mete out punishment, to all peace efforts and all war-mongering — prior to that other spirit which would establish and secure peace, *pax*, by pacts. The space of this freedom from revenge is prior to all pacifism, and equally to all power politics (CT, 88).

Nietzsche often spoke of contemporary man as "the last man" — suggesting, and perhaps updating, man in Hesiod's disastrous "Iron Age." The "last man," in order to create himself beyond the man of revenge, to what Nietzsche refers to as the "overman," must of his own accord, to the extent possible, open himself to a new relation to nature, to life, and to the essential, original relationship — that between the sexes. Man — with "the spirit of revenge as the subject of his best reflection" is a *bridge* to man beyond himself. Nietzsche's view appears to strongly suggest that without this "deliverance from revenge" the future of humanity may well be in jeopardy. Like a disease, unchecked, the poison continues spreading throughout the human community. Nietzsche did perceive what he believed was a grave and imminent danger.

The ideas which Nietzsche spent most of his life developing, thinking and rethinking, relating, arranging and rearranging, were the ideas which served him in his critique of Christianity — his tools and weapons. The major ones were nature, life, power, values, bodies, sexuality, conflict or competition, psychology — especially resentment, revenge, and fabrication. Always he reminded us that what he had to say was based upon his experiences, his perspectives, his interpretations, his "truths."

There is every reason to assume that Nietzsche would have assembled these same ideas in forging an assault against hostile, violent war. I believe he would have applied them similarly, and that he would have arrived at similar conclusions. War, like Christianity, had played a dominant role in effectively bringing about the degradation and devastation of Western culture.

Reflecting the interests and questions of Einstein in the early twentieth century, this book endorsed the question, or problem, of "Why War?" And, as Einstein suggested, this question appears to solicit a reply from a psychological perspective or perspectives. Which is well within the purview of Nietzsche's thinking. He surely would have considered the query as looking for the origin, or source, of war within the psyche.

But considered as meaning originating in the psyche, Nietzsche typically would have added the question, or questions, pointing to "who?" Whose psyche, or psyches? War involves many persons. Who invents war? Who dies? Who kills? Who survives? Who grieves? Who wins and who loses? Who sacrifices? Who is the enemy? And many more questions of identity. Recall briefly this explicit reference in *Beyond Good and Evil*, when Nietzsche writes:

> Gradually I have come to realize what every great philosophy up to now has been: the personal confession of its originator, a type of involuntary and unaware memoirs; also that the moral (or amoral) intentions of each philosophy constitute the protoplasm from which each entire plant has grown. Indeed, one will do well

(and wisely), if one wishes to explain to himself how on earth the more remote metaphysical assertions of a philosopher ever arose, to ask each time: What sort of morality is this (is *he*) aiming at? Thus I do not believe that a "desire for comprehension" is the father of philosophy, but rather a quite different desire has here as elsewhere used comprehension (together with miscomprehension) as tools to serve its own ends. Anyone who looks at the basic desires of man with a view to finding out how well they have played their part in precisely this field as inspirational genii (or demons or hobgoblins) will note that they have all philosophized at one time or another... Conversely, there is nothing impersonal whatever in the philosopher. And particularly his morality testifies decidedly and decisively as to *who he is* — that is, what order of rank the innermost desires of his nature occupy (*BGE*, 6, 7).

A few pages later Nietzsche carefully defines what he is meaning by "morality."

... morality understood as the doctrine of the rank-relations that produce the phenomenon we call "life." — (*BGE*, 22).

The question has been raised before — can the producers of the "phenomenon we call life" be understood in any way more fundamental than referring to female and male? And, great philosophers — *who* have they been? *Mostly males.* Best early examples — Plato and Aristotle — inventing value systems which ranked the powers, or abilities, of the male as greater than the powers, or abilities, of the female.

Much the same "innermost desires of his nature" were revealed when Nietzsche addressed the question to theologians and priests, or "priestly types." They, too, had invented a set of religious ideas, the aim of which was to create an order of power, of value, of rank that unsurprisingly placed themselves as the privileged ones. And *who* were they? *Mostly males* — creators of Christian morality, the morality Nietzsche refers to as "slave-morality." And these priestly types? Nietzsche refers to them as "decadents." He credited male philosophers and male theologians — (best example of the latter being Thomas Aquinas) with having been the chief producers of values. This was their desire and aim, the fulfillment of which Nietzsche believed had been decidedly successful.

When Nietzsche, very late in his creative life, wrote *Ecce Homo*, the opening sentence of his Foreword was this:

SEEING that I must shortly approach mankind with the heaviest demand that has ever been made on it, it seems to me indispensable to say *who I am*. This ought really to be known already: for I have not neglected to "bear witness" about myself. But the disparity between the greatness of my task and the *smallness* of my contemporaries has found expression in the fact that I have been neither heard nor even so much as seen. I live on my own credit, it is perhaps merely a prejudice that I am alive at all? . . I need only to talk with any of the "cultured people" who come to the Ober-Engadin in the summer to convince myself that I am not alive. . . Under these circumstances there exists a duty against which my habit, even more the pride of my instincts revolts, namely to say: *Listen to me! For I am thus and thus. Do not, above all, confound me with what I am not!* (EH, 33)

This book, *Ecce Homo*, is Nietzsche's reflection on *who he is* and what he has become, and for a brief time is still becoming. It is his "truth," among the most beautifully written of world-historical memoirs.

Nietzsche's attraction to thinking, or rethinking, *nature*, is reflected in his early interest in Hesiod, described as "the poet of the roadside grass and the many colored earth, and of men who live by the soil. Echoes of ancient peasant wisdom, and of the mysteries of the earth. . ." Nietzsche was well acquainted with the pre-Socratics and their particular interpretations of physical nature, and he remained an admirer of Heraclitus. Nietzsche knew well the rational and logical approach of Aristotle. Nature was divided into animate and inanimate "substances," with two different and unequal animate forms — male and female. After centuries of religious adaptations of Aristotelian analyses, Hobbes entered the discussion of nature with the adoption of a new scientific approach. And he spoke of "natural man" and of an original condition of humans, which he referred to as the "state of nature." However, in reading and rereading Nietzsche, the evidence grows that the most important influence on his centering himself back into the natural world was Bachofen. I am convinced that Bachofen's approach to cultural history, and his opening up the wealth of new opportunities to understanding our species through myths and symbols was exciting and inspiring for Nietzsche. Bachofen himself wrote, "What I am engaged upon now is the true study of nature."

Remember, Bachofen said this. From the earliest sources the history of Western culture provides abundant evidence that this history has as its basic guiding "law," a *formative pattern*, of two natural, original forms — one form female the other form male — evolving in a process of continual conflict, constant changes in the powers, the meanings of those powers, and the evaluations of those powers. According to Bachofen, this formative pattern had developed from the earliest stages of the maternal — the primacy of the mother/child relationship — to the later stages of the dominance of the paternal. Religion, which reflected these changing powers and values, was the vehicle, the bearer, the primary force which, Bachofen believed, shaped any culture. And although he frequently sounds as if he prefers the earlier, ancient pattern, nevertheless he seems compelled to believe that later paternal stages were representative of higher stages of cultural evolution. Patriarchal religions were vastly different — and significantly more advanced — than earlier matriarchal religions.

I am convinced that Nietzsche embraced from Bachofen the opportunities for exciting new possibilities in the direction his own thinking was moving already, early in its development. Bachofen's interpretation of the process of cultural evolution presented Nietzsche with the basis of his own interpretation of cultural history. Taking a hard look at religions of his own time, especially Christianity — the dominant religion of the West — his critical perspectives, and the importance of the idea of contrariety or opposition, Nietzsche *reversed* the view of Bachofen. The evolution of the paternal pattern of Western culture had really been a process of "devaluation." Nietzsche increasingly intensified his description of German culture, and Western culture, as degenerating, deteriorating, "a wasteland." The hints and clues that he offers are to be found throughout the entire body of his works. What follows is one small sample. Very late in *Ecce Homo*, he wrote:

> I now have the skill and knowledge to *invert perspectives*; first reason why a "revaluation of values" is perhaps possible at all for me alone (*EH*, 40).

Nietzsche's insistence on pressing his idea of perspectivism — of indefinite possibilities of perspectives — took him to this position of being able to "invert perspectives." Two critical instances of this ability, or power, were one — to reverse the order of Bachofen's interpretation of history — and two — to see the world from the perspectives of the female, in addition to his own male perspectives. To view the world from the position of "upside down." And he relished and developed this power!

This is an opportunity to call again on Hesiod. Hesiod, the poet, an art especially dear to Nietzsche — who himself wrote poetry. But Hesiod also wrote history — in the form of poetry. For many important inspirations, Nietzsche was indebted to Bachofen. History — the importance of studying history "scientifically" — and the notion that to think philosophically should be understood as philosophy becoming both historical and scientific. And in his philosophizing historically, Nietzsche was drawn frequently, and perhaps more forcefully back to Hesiod. It is worthwhile to include here a more lengthy portion of Hesiod's *Works and Days*:

> *Or if you will, I will outline it for you*
> *in a different story,*
> *well and knowledgeably — store it up*
> *in your understanding —*
> *the beginnings of things, which were the same for gods*
> *as for mortals.*
> *In the beginning, the immortals*
> *who have their homes on Olympos*
> *created the golden generation of mortal people.*
> *These lived in Kronos' time, when he*
> *Was the king in heaven.*
> *They lived as if they were gods,*
> *their hearts free from all sorrow,*
> *by themselves, and without hard work or pain;*
> *no miserable*
> *old age came their way; their hands, their feet,*
> *did not alter.*
> *They took their pleasure in festivals,*
> *And lived without troubles.*
> *When they died, it was as if they fell asleep.*
> *All goods*

were theirs. The fruitful grainland
yielded its harvest to them
of its own accord; this was great and abundant,
while they at their pleasure
quietly looked after their works,
in the midst of good things
[prosperous in flocks, on friendly terms
with the blessed immortals.]
Now that the earth has gathered over this generation,
these are called pure and blessed spirits;
they live upon earth,
and are good, they watch over mortal men
and defend them from evil;
they keep watch over lawsuits and hard dealings;
they mantle
themselves in dark mist
and wander all over the country;
they bestow wealth; for this right
as of kings was given them.
Next after these the dwellers upon Olympos created
a second generation, of silver, far worse
than the other.
They were not like the golden ones either in shape
or spirit.
A child was a child for a hundred years,
looked after and playing
by his gracious mother, kept at home,
a complete booby.
But when it came time for them to grow up
and gain full measure,
they lived for only a poor short time;
by their own foolishness
they had troubles, for they were not able

to keep away from
reckless crime against each other,
nor would they worship
the gods, nor do sacrifice on the sacred altars
of the blessed ones,
which is the right thing among the customs of men,
and therefore
Zeus, son of Kronos, in anger engulfed them,
for they paid no due
honors to the blessed gods who live on Olympos.
But when the earth had gathered over this generation
also — and they too are called blessed spirits
by men, though under
the ground, and secondary, but still
they have their due worship —
then Zeus the father created the third generation
of mortals,
the age of bronze. They were not like
the generation of silver.
They came from ash spears. They were terrible
and strong, and the ghastly
action of Ares was theirs, and violence.
They ate no bread,
but maintained an indomitable and adamantine spirit.
None could come near them; their strength was big,
and from their shoulders
the arms grew irresistible on their ponderous bodies.
The weapons of these men were bronze,
of bronze their houses,
and they worked as bronzesmiths. There was not yet
any black iron.
Yet even these, destroyed beneath the hands
of each other,

went down into the moldering domain of cold Hades;
nameless; for all they were formidable black death
seized them, and they had to forsake
  the shining sunlight.
Now when the earth had gathered over this generation
also, Zeus, son of Kronos, created yet another
fourth generation on the fertile earth,
  and these were better and nobler,
the wonderful generation of hero-men, who are also
called half-gods, the generation before our own
  on this vast earth.
But of these too, evil war and the terrible carnage
took some; some by seven-gated Thebes
  in the land of Kadmos
as they fought together over the flocks of Oidipous;
  others
war had taken in ships over the great gulf
  of the sea,
where they also fought for the sake
  of lovely-haired Helen.
There, for these, the end of death was misted
  about them.
But on others Zeus, son of Kronos, settled a living
  and a country
of their own, apart from human kind,
  at the end of the world.
And there they have their dwelling place,
  and hearts free of sorrow
in the islands of the blessed
  by the deep-swirling stream of the ocean,
prospering heroes, on whom in every year
  three times over
the fruitful grainland bestows its sweet yield.

These live
far from the immortals, and Kronos
is king among them.
For Zeus, father of gods and mortals,
set him free from his bondage,
although the position and the glory still belong
to the young gods.
After this, Zeus of the wide brows
established yet one more
generation of men, the fifth, to be
on the fertile earth.
And I wish that I were not any part
of the fifth generation
of men, but had died before it came,
or been born afterward.
For here now is the age of iron. Never by daytime
will there be an end to hard work and pain,
nor in the night
to weariness, when the gods will send anxieties
to trouble us.
Yet here also there will be some good things
mixed with the evils.
But Zeus will destroy this generation of mortals
also,
in the time when children, as they are born,
grow gray on the temples,
when the father no longer agrees with the children,
nor children with their father,
when guest is no longer one with host,
nor companion to companion,
when your brother is no longer your friend,
as he was in the old days.
Men will deprive their parents of all rights,

as they grow old,

and people will mock them too,

babbling bitter words against them,

harshly, and without shame in the sight of the gods;

not even

to their aging parents will they give back

what once was given.

Strong of hand, one man shall seek

the city of another.

There will be no favor for the man

who keeps his oath, for the righteous

and the good man, rather men shall give their praise

to violence

and the doer of evil. Right will be in the arm.

Shame will

not be. The vile man will crowd his better out,

and attack him

with twisted accusations and swear an oath

to his story.

The spirit of Envy, with grim face

and screaming voice, who delights

in evil, will be the constant companion

of wretched humanity,

and at last Nemesis and Aidos, Decency and Respect,

shrouding

their bright forms in pale mantles, shall go

from the wide-wayed

earth back on their way to Olympos,

forsaking the whole race

of mortal men, and all that will be left by them

to mankind

will be wretched pain. And there shall be no defense

against evil (H, 3l, 33, 35, 37, 39, 4l, 43).

As with Nietzsche's writings, it is important to listen carefully to each word of Hesiod. But there is a clear process of development in Hesiod, of Ages — from gold to iron. In Hesiod's account, a cataclysmic descent, in the poetic imagery or symbolism of metals. Gold — precious, malleable, ductile, and the corresponding description of the Golden Age. The superior era and the superior metal. Devolving to his, and our age — Iron, and the corresponding Iron Age. The lowest base metal, suggesting uses of strength and violence, and the most debased age, or culture. From superior age to the most inferior.

Nietzsche certainly interpreted the direction of cultural history as a process, not ascending from inferior to superior stages, but as a process of descending from ancient, superior stages to inferior stages. He adopted the model of Hesiod's poetic epic. Both Hesiod and Nietzsche viewed the present age, or culture, as one characterized by depravity — Hesiod in the seventh or eighth century BC and Nietzsche in the nineteenth century. Hesiod interpreted the source of this corruption as envy. Nietzsche's interpretation centered on resentment. The process had been that of continuous transformation of values — of devaluing.

There is another short passage from Hesiod's poem that Nietzsche surely would have found compelling. True, I think Bachofen had much influence on Nietzsche's fascination with the endless possibilities which existed in looking back into ancient history on the use and power of symbols and images. But Nietzsche was surely aware of this short story of Hesiod's:

> Now I will tell you a fable for the barons;
>   they understand it.
> This is what the hawk said when he had caught
>   a nightingale
> with spangled neck in his claws and carried her
>   high among the clouds.
> She, spitted on the clawhooks, was wailing pitifully,
>   but the hawk, in his masterful manner,
>   gave her an answer:
> "What is the matter with you? Why scream?

Your master has you.

You shall go wherever I take you,

for all your singing.

If I like, I can let you go. If I like,

I can eat you for dinner.

He is a fool who tries to match his strength

with the stronger.

He will lose his battle, and with the shame

will be hurt also."

So spoke the hawk, the bird who flies so fast

on his long wings.

But as for you, Perses, listen to justice;

do not try to practice

violence; violence is bad for a weak man; even a noble

cannot lightly carry the burden of her,

but she weighs him down

when he loses his way in delusions; that other road

is the better

which leads toward just dealings. For justice

wins over violence

as they come out in the end. The fool knows

after he's suffered (H, 43, 45).

Very early in his "Homer's Contest", Nietzsche wrote this:

> When one speaks of *humanity*, the idea is fundamental that this is something which separates and distinguishes man from nature. In reality, however, there is no such separation: "natural" qualities and those called truly "human" are inseparably grown together. Man, in his highest and noblest capacities, is wholly nature and embodies its uncanny dual character (*PN*, 32).

What, if not the uncanniness of two forms of life, the original, enduring, recurring pattern of the sexes? And this in Nietzsche's first major work, *Human, All Too Human*:

> Now everything *essential* in human development occurred in pri-
> meval times, long before those four thousand years with which
> we are more or less familiar (*HA*, 14).

Nietzsche saw clearly that philosophy must become historical, and
that historical philosophizing would never even consider not recognizing as central to history the dynamics of power between males and females. Another reminder:

> *A male culture.* Greek culture of the Classical era is a male culture.
> As for women, Pericles, in his funeral oration, says everything
> with the words: "They are best when men speak about them as
> little as possible" (*HA*, 156, 157).

Eight years after the publication of *Human, All Too Human*, Nietzsche
(as he frequently did for other of his books) wrote a Preface in which
he reflected:

> Whoever guesses something of the consequences of any deep
> suspicion, something of the chills and fears stemming from isola-
> tion, to which every man burdened with an unconditional *differ-
> ence of viewpoint* is condemned, this person will understand how
> often I tried to take shelter somewhere, to recover from myself,
> as if to forget myself entirely for a time (in some sort of reverence,
> or enmity, or scholarliness, or frivolity, or stupidity); and he will
> also understand why, when I could not find what I *needed*, I had
> to gain it by force artificially, to counterfeit it, or create it poeti-
> cally (*HA*, 4).

Nietzsche not only adopted as some of his own the perspectives of
the woman, he added this:

> The greatest distinction that fate can bestow on us is to let us
> fight for a time on the side of our opponents. With that we are
> *predestined* for a great victory (*GS*, 255).

Later in his productive life, in what he referred to as "the eternal
war between the sexes," it could be said that Nietzsche was fighting
on both sides of the war. Both males and females were losers in the
devaluation that had taken place within Christianity. In other words,
Nietzsche's function, his "fate," was to liberate both males and females
— to return to, to reaffirm, the values of antiquity. He did foresee the
overturning of the values of his own time and the necessity for a "revaluation of all values." Becoming centered on the power of values to
shape a culture, and the power of those individuals whose power it is to

create values — and thereby shape a culture — Nietzsche's target was the *value of persons*, and these were moral values.

Whatever interpretation he thought best addressed the questions of the ordering of sexual powers and values in the early stages of human history, Nietzsche clearly praised and preferred the table of values which he, like Bachofen, believed had been revealed in myths and symbols. The ancients affirmed the creative powers of nature, of life, of the body, of sexuality, of procreation, with special emphasis and acclaim for the natural creative power — and therefore the value — of the woman as giving birth to new life, to the *child*. Maternal power was preeminent. Nietzsche, as did Bachofen, interpreted religious ceremonies and practices as reflecting correctly these social and cultural values. Recall these words:

> What is astonishing about the religiosity of the ancient Greeks is the lavish abundance of gratitude that radiates from it. Only a very distinguished type of human being stands in *that* relation to nature and to life. Later, when the rabble came to rule in Greece, *fear* choked out religion and prepared the way for Christianity (*BGE*, 58).

The original values of antiquity had been devalued, inverted, reversed, denied, transformed in the Christian religion — to devalue nature, life, the body, sexuality, procreation, the maternal power of giving birth. Paternal power, or powers, became dominant, and the power and value of the woman declined. Nietzsche, I believe, was saying that with significant evidence, this earliest order or pattern of values was "natural," and the reversed pattern "unnatural." And he would somehow prefer a return to those values as expressed in the Dionysian religion. He "became" Dionysus, a god favored and worshipped by women. A kind of "war between the sexes" — competition, conflict (perhaps tension might be a better term) — was a natural pattern of expression of powers. In *Ecce Homo*, when Nietzsche is reflecting on his earlier writing in *Beyond Good and Evil*, he wrote this:

> The task for the immediately following years was as clear as it could be. Now that the affirmative part of my task was done, it was the turn of the denying, the No-saying and *No-doing* part — the revaluation of existing values themselves, the great war — the evocation of a day of decision. Included here is the slow search for those related to me, for such as out of strength would offer me their hand for *the work of destruction*. — From now on all

> my writings are fish-hooks: perhaps I understand fishing as well
> as anyone?... If nothing got *caught* I am not to blame. *There were no*
> *fish...* (*EH*, 112).

But this was related to the "great war." And war — hostile, violent, deadly, devastating — that had preceded him for centuries, was
rampant during his own time, and grew a few years later to monstrous
proportions he could never have foreseen. Nietzsche abhorred German
nationalism, Christianity, and anti-Semitism. And he believed that the
ascendancy of military values was simultaneously accompanied by
the decline of cultural values. The deteriorating conditions of German
culture, European culture, and Western culture were consequences of
religious values and *military values*. So, we are looking at *war and the values*
of war. Nietzsche saw both Christianity and hostile, violent, deadly war
as antithetical to culture — in direct and unequivocal opposition. Also,
as antithetical to nature. Religion and war — together, a formidable
pair of powers!

So, the questions. First, *what* is war? Or, perhaps more accurately,
what is called "war"? It is an exercise of power, or powers — power
understood as control or authority; as efforts to obtain or maintain a
position of power by establishing a ranking order of persons and values. War is not about land, oil, dollars, tanks, guns, aircraft, weapons of
mass destruction. War is about people, persons persuaded to believe,
trained to kill, to die if necessary. It is about human bodies damaged
and suffering — lost.

Second, *who* is involved in war? Speaking generally, members of
both bodily forms — males and females. Of every possible age, size,
race, nationality, religious persuasion, political position, economic or
educational condition, directly and indirectly. Speaking specifically,
war is invented and marketed by a relatively small clique, *mostly males*.
Speaking specifically again, war is engaged in by, and directed toward,
children — some mother's child — *mostly males*.

And third, *why* war? Or, perhaps closer to Nietzsche's interests,
what is the *psychological origin*, or source of war? And, as was his answer
regarding Christianity, no reason not to believe it would have been the
same answer — *resentment* and desire for *revenge*. Resentment understood as "something perceived and interpreted as an insult or injury"

— arising from a perceived *absence* of power of the inventors of war and directed toward the *presence* of power of the female — the child-bearing mother.

And then fourth, *how* is war — hostile, violent war — accomplished repeatedly, or maintained? What are the tactics? These stand out — fear, horror, torture, ignorance, taboos, exclusion, obedience, secrecy, silence, and especially *lies* and *deception*. As with his unmasking of the real nature of Christianity, so it would be, or have become, with Nietzsche's unmasking of war. It is a *lie*, a fabrication engendered to transform what is incomprehensibly vile and vicious into its opposite — something virtuous, glorious, heroic, necessary, justified — a brilliant act of alchemy. And generating ever more lies to assist, extend, and cover up the source — resentment and the desire for revenge. And as Nietzsche says — dissatisfaction with themselves, their bodies. "Despisers of their bodies."

Perhaps Nietzsche would have said, or would say, something like this. In determined and persistent efforts to reorder the ancient, perhaps original, perhaps natural, patterns of values and of authority, power — between females and males — selected groups of males had been creating and recreating elaborate plans, or strategies. This process of revaluing, or devaluing, had not only become a failure and ultimately unsuccessful, but had become, and was becoming, rapidly and increasingly disastrous and terrifying.

Consider further — perhaps this from Nietzsche. In the drive by these males for acquiring, maintaining, and increasing their power — and specifically their power, and thereby value, in relation to the power and value of the female — the strategic reasoning continued being developed in two distinct versions, two reinforcing, complementary approaches. If we listen carefully to Nietzsche, the religion of Christianity had carefully provided and established the ideas, the ideology, the narrative within which the ranking order of values had been the center, the primary goal. Nietzsche referred to these as the "doctrine of morality" — moral values. War, provided the execution, or implementation, of these ideas — to maintain and enhance the same order of power and value, the same desire for control. And both Christianity and war had essentially the same psychological origin. Different fields of operation, different modes of influence, different narratives. *Together*, they were laying waste Western culture — "the wasteland grows."

Christianity and war — different narratives, but sharing many ideas or words, overlapping and reinforcing. Some important ones — sacrifice, command, obedience, reward, punishment, duty, disgrace, honor, glory, victory, defeat, praise, weapons, battle, guilt, danger, bodies. And there are others, perhaps less obvious. Both Christianity and war are infused with ranking, hierarchical orders — of powers and values — and are *mostly male*. Perhaps this manner, or method, of the ordering of values is more immediately visible in the military than in religious institutions. Perhaps not.

And what would Nietzsche say of these two narratives? That both were, and are, a *fabrication*. Both narratives "invented, created, made up for the purpose of deception, specifically constructed from diverse and usually standardized parts." Fabrication has been the instrument, the necessary implement, for designing and structuring a value system which has proved to be a disaster.

Back to basics — Nietzsche's basics. The nature of this universe, this world, in which humans participate, is best understood as a "monster of energy." As a participant, each living being — human beings in particular — are basically and necessarily driven by an impulse called the "will to power." Life is best understood as a process of seeking, of striving for, more power, the "feeling of power" — of becoming powerful, or more powerful — power understood as *ability* to act or produce. Happiness, according to Nietzsche, is "the feeling that power is increasing."

Life is present in nature in two forms — female bodies and male bodies — the prototypical *pair of contraries*. Similar and different, a unity, interdependent — with different experiences, different perspectives, different interpretations, different abilities or powers, different evaluations. Each is conscious of oneself and then of the other — the primary, essential "other." Each one has its individual, personal, unique history, its own process of becoming, of creating oneself. All persons are participants of the process of natural history and of cultural history. The *formative pattern* of all human histories — natural, personal, and cultural — is the continuing dynamics of power between males and females, and the respective creative powers of each.

The power of creating life, the power of creating oneself, and the power of creating values. The most important values are those signifying the value of individual persons, according to their powers, i.e., their abilities. This evaluating of abilities — what Nietzsche means by morality — is the basis of every culture.

Interpretations of power, of sexuality, of history, of values are expressed and implemented through the *power of words* — the signifying power of speech. The ability to name or change names; to create definitions and meanings; to question and suggest answers; to identify problems and experiment with solutions; to offer hypotheses; to create fabrications — and much more. To speak and to silence others, to restrain others from exercising the power of speaking. Nietzsche understood well, and the power of his words was exhilarating — for him and for us.

Nietzsche placed his bets on the opportunities which might develop within the newly emerging science of psychology for greater understanding of human behavior. For him, becoming "psychological" entailed these aspects — experiential, experimental, historical, physiological, and philosophical. And his significant psychological discovery, deeply concealed within the psyche, was resentment, with the desire for revenge. This psychological phenomenon became the key with which he laid bare what he believed were the destructive fabrications of Christianity.

Nietzsche recognized, and understood well, the various possible expressions of this will to power, this constant drive for the "feeling of power." There is no doubt that for him, power is best, and first, to be thought of as *ability* — the power to act, to produce, to create. In addition to ability is the sense of power as *dominance*, and the several terms suggesting dominance — control, rule, authority, command, force, might, strength. The constant process of the dynamics of power between females and males is constantly shifting in the emphasis, or emphases, of these different meanings — especially the two, power as ability and power as dominance. And resentment figures as a major factor in this process. Nietzsche demanded that we pay attention. Resentment — a feeling reflecting a lack, an absence of ability, an inability.

As we know, these basic ideas required the context of history — of historical development. Nietzsche's determination to "think historically" and his own interpretations of cultural history found their sources particularly in the thinking of Hesiod, of Hegel, and of Bachofen. Of course, Nietzsche affirmed completely within these interpretations the notion of Heraclitus' Logos — the philosophical principle that governs and develops the universe — the principle of constantly interacting opposites. Hesiod thought of history as a process developing from the earliest stage of the past, which he described as golden, to his present era, which he described as the iron age. From his perspective, man's future appeared bleak — but with possibilities for renewal. The direction of history thus far had been one of catastrophic decline.

Hegel's views of history were from a different perspective, in a different time, but had a profound influence on Nietzsche's views — as did those of Hesiod. Hegel considered human history as the process of the development of the consciousness of freedom — revealed as a process involving first consciousness of the self, acting to realize its own potentialities, followed by consciousness of the "other." A unity of opposites in their opposition (very Heraclitean). History appeared as a creative process of opposing points of view seeking reconciliation — the demonstration of the underlying compatibility of things that seem to be incompatible, or to restore harmony or friendship. Nietzsche's entire philosophy, in one sense, is his engagement with the logical opposites — affirmation and negation — emphasized by Hegel.

Bachofen's interpretation of cultural history appears to have been significantly influenced in part by Hegel. However, his own adaptation relied mostly on his study of myths and symbols, and he also gave a positive, progressive interpretation. From what he calls matriarchy and the important relationship of the mother and child, to patriarchy, and the primacy of the father. From natural religion tied closely to nature (which he refers to as Dionysian religion) to supernatural religion. From the early dominance of the maternal principle and female perspective, to the later paternal principle and male perspective.

We certainly know how important history, and historical consciousness, was to become for Nietzsche. And his own historical consciousness reflected these other interpretations of history, which he nevertheless rethought, and to which he added new imaginative key phenomena. Nietzsche's interpretation of cultural history became much richer, more comprehensive, and much more combustible. Generally, cultural history had been a creative process, changing perhaps in some ways which warranted calling it "cyclical," or "eternally recurring."

"Thinking historically" in conventional terms of past, present, and future, Nietzsche started with the present, reached far into the past for understanding of the present, and for some dim, uncertain vision of the future. His experiences of the present Western culture, as he stressed emphatically, led to his characterization of European culture as "beautiful ruins" — degenerating, deteriorating, dominated by Christianity and religious values. What follows is an adventurous compilation of multiple aspects of his interpretation, clearly reflecting his engagement with the principle of contrariety and his claim to have the ability to "invert perspectives," and thereby to possibly be in a position for a "revaluation of all values."

Cultural history had moved from higher stages to lower stages; from affirmation of life to negation of life; from natural order to cultural chaos; from power meaning ability to power meaning dominance; from participation in nature to dominance of nature; from natural religion to supernatural religion; from natural casual conflict (of abilities) to unnatural hostile war (of dominance); from truth to untruth; from ability of self to non-ability of self; from valuation to devaluation; from "master-morality" to "slave-morality"; from the feminine principle to the masculine principle; from the primacy of the maternal relationship to the primacy of the paternal relationship; from female values to male values; from the voice of the female to the voice of the male (and female silence.)

As to Nietzsche's image, or images, of the future, the following stand out — some explicit, others typically more concealed, often delivered metaphorically or symbolically. The changes emerging would be an "earthquake," an era of "grand politics," a "revaluation of all values," a further evolution of human consciousness of the eternal dynamics of power between males and females. The future would, or could, bring freedom from resentment and revenge, especially for the male, and freedom from violence, especially for the female, the violence which is frequently generated by resentment — a reconciliation. Of his several ways of giving his version of the man he was anticipating in the future, nothing Nietzsche says is more powerful than this aphorism written far along in his development:

> Not the strength but the permanence of superior sensibilities is the mark of the superior man (*BGE*, 74).

But the most unexpected, overlooked, perhaps amazing perspective on the future was his "return" to focus on the *child*. Both literally and symbolically, the child embodies and represents the future of the human species. As Nietzsche saw, and said, in any revaluation, the highest value will, or must, be given to the child. Man's very future depends upon it — as he reminded us. Perhaps Nietzsche was speaking from his "inverted perspective," which was, I believe, his feminine perspective. And speaking in his "inverted voice," that of the woman, recalling early times in history. In any case, he was suggesting, I believe, that the creative abilities of both females and males could be developed toward affirming life — one's own, and especially that of the child — to hear, to care for, to protect, to defend, to enhance — to insure the future.

In *Human, All Too Human*, Nietzsche wrote a short chapter entitled "Woman and Child." Later, in what he considered his best and most important book, *Thus Spoke Zarathustra*, the first speech of Zarathustra, "On the Three Metamorphoses" reads:

> Of three metamorphoses of the spirit I tell you: how the spirit becomes a camel; and the camel, a lion; and the lion, finally, a child. . .

> But say, my brothers, what can the child do that even the lion could not do? Why must the preying lion still become a child? The child is innocence and forgetting, a new beginning, a game, a self-propelled wheel, a first movement, a sacred "Yes." For the game of creation, my brothers, a sacred "Yes" is needed: the spirit now wills his own will, and he who had been lost to the world now conquers his own world (*Z*, 25, 26, 27).

In the final speech of Zarathustra, Nietzsche wrote:

> About all this Zarathustra spoke but a single sentence: "*My children are near, my children.* Then he became entirely silent (*Z*, 326).

References to the child, or to children, are a special theme throughout *Thus Spoke Zarathustra*. The child as metaphor, or symbol of creativity. And Christianity had devalued the mother and the child, not as symbol, but in nature, in life. Zarathustra's affirmation is one of ecstasy. For Nietzsche, to reaffirm the value of nature, of life, of the body, of sexuality, meant above all reaffirming the value of the child. When he voiced his alarm regarding the future and its relationship to values, particularly to the values of Christianity, he wrote:

> The "real world" and the "apparent world" — in plain terms: the *fabricated* world and reality. . .The *lie* of the ideal has hitherto been the curse on reality, through it mankind itself has become mendacious and false down to its deepest instincts — to the point of worshipping the *inverse* values to those which alone could guarantee it prosperity, future, the exalted *right* to a future (*EH*, 34).

Nietzsche made clear that his first major task was to expose the destructive nature of the values of Christianity. In any preliminary stage of the process of revaluation — the next task that he set for himself — would he not have directed his attention to the obvious destructive nature of the values of violent, vile, unnatural war? He had voiced his awareness of the contrary relationship between war and culture — of the ruinous effects of war. That was obvious in nineteenth century Europe, and confirmed by many examples in history. However, he had never explicitly made war the target of an exhaustive critique as he had done with Christianity.

Were we able to ask Nietzsche for his views on war, for a critique of war, similar perhaps, to his comprehensive critique of Christianity, he would no doubt have replied that whatever understanding he might have would be based on his own *experiences*, and that perhaps we should learn to listen more carefully to his words — as he suggested, "to hear him with our eyes." And we are trying.

Images, symbols, words, and people spoke to Nietzsche, and he listened. Then he thought and rethought what he had received and passed these on to us. Listen again to these words written late — "the eternal *war* between the sexes." These simple, deceptively trivial words captured the essential, most significant complexity of human history.

The earliest accessible images in nature, and throughout human history, seemed to reveal in concert some natural tension between females and males. A wealth of these earliest images, and symbols, and words developed one central theme or cultural pattern around the female expressing the perspectives, interpretations, voices, the natural "truths," the natural "rights," the powers and values of the female. That is the maternal principle, meaning the predominance of the creative powers of the woman, and especially of the mother–child relationship as one of *belonging* — meaning bound by birth, allegiance, or dependency — close, intimate, embodying the continuity of life. This relationship appeared to favor the mother–daughter connection as closely representing, and being part of, the natural, creative, life process. And this preeminence was sanctified in the natural religiosity of the ancients — *mostly female.*

The process of reversal, of denial, of negation had been one with a very long history. It had been, however, revolutionary in its accomplishments. A cultural pattern, or theme, developed around the male, replacing the female pattern with that of the male. Perspectives, interpretations, voices, "truths," "rights," powers, values — from the maternal to the paternal, from the primacy of the mother/child (daughter) relationship, to that of the father/child (son). And this preeminence became sanctified in patriarchal, supernatural religions.

Nietzsche, as did Bachofen, viewed this transformation as the most important turning point in the history of the relations between the sexes. And therefore for Nietzsche, in the history of human culture.

Two major shifts, perhaps overlooked during the long transformation. One was the emphasis very early on the sense, or meaning, of the phenomenon of power. In the maternal culture, power appears to have been signifying *ability,* or *abilities.* In the later paternal culture, the notion of power appears to have focused attention on the meaning of *dominance,* or *control.* Failure to recognize this discrepancy, to uncritically assume that the meaning of dominance is, or had been, the preferred and defining sense, leads to muddled interpretations of Nietzsche's preferences of meaning.

The second, more subtle, shift — interwoven with the first — were the sensibilities concerned with the child. As suggested, the relationship of the mother and child was one of *belonging* — bound by birth, allegiance, dependency — close and intimate. The relationship of the father and child became more one of *possession* — of ownership, of property. From an earlier link between *abilities* and *belonging* to a link between *dominance* and *possession.* The magnitude of the difference, or change, in the interpretation of the relative position and value of the child, and on the transmission of cultural values to future generations, cannot be overstressed.

In *Human, All Too Human*, Section Five, entitled "Signs of Higher and Lower Culture," Nietzsche wrote:

> *A male culture.* Greek culture of the Classical era is a male culture. As for women, Pericles, in his funeral oration, says everything with the words: "They are best when men speak about them as little as possible."

> The erotic relationship of men to youths was, on a level which we cannot grasp, the necessary, sole prerequisite of all male education (more or less in the way love affairs and marriage were for a long time the only way to bring about the higher education of women); the whole idealism of strength of the Greek character was thrown into that relationship, and the treatment of young people has probably never again been so aware, loving, so thoroughly geared to their excellence (*virtus*), as it was in the fifth and sixth centuries — in accordance with Holderlin's beautiful line, "denn liebend giebt der Sterbliche vom Besten" (for loving the mortal gives of his best). The more important this relationship was considered, the lower sank interaction with women: the perspective of procreation and lust — nothing further came into consideration; there was no spiritual intercourse with them, not even a real romance. If one considers further that woman herself was excluded from all kinds of competitions and spectacles, then the sole higher entertainment remaining to her was religious worship.

> To be sure, when Electra and Antigone were portrayed in tragedies, the Greeks *tolerated* it in art, although they did not like it in life; just as we now do not tolerate anything with pathos in *life*, but like to see it in art.

> Women had no task other than to produce beautiful, powerful bodies, in which the character of the father lived on as intact as possible, and thus to counteract the increasing overstimulation of nerves in such a highly developed culture. This kept Greek culture young for such a relatively long time. For in Greek mothers, the Greek genius returned again and again to nature (*HA*, 156, 157).

Immediately following this quotation, in the same section, Nietzsche wrote:

> *Prejudice in favor of size.* Men clearly overestimate everything large and obtrusive. This came from their conscious or unconscious insight that it is very useful if someone throws all his strength into one area, and makes of himself, so to speak, one monstrous or-

gan. Surely, for man himself, a *uniform* cultivation of his strengths is more useful and beneficial, for every talent is a vampire that sucks blood and strength out of the remaining strengths; and excessive productivity can bring the most gifted man to madness. Even within the arts, extreme natures attract notice much too much, but a much lesser culture is also necessary to let itself be captivated by them. Men submit from habit to anything that wants to have power (*HA*, 157, 158).

Nietzsche's German culture of the nineteenth century was also a male culture — determinably authoritarian, increasingly exemplifying force, strength, physical might, physical violence, increasingly militarized — commanding and obeying. And in Nietzsche's view, attending less to creative abilities — especially those of the female.

Contemporary Western culture, and most global cultures, remain male cultures — or *mostly male.* Exemplars — politics, sports, industry, religion, business, military, entertainment, arts, sciences, philosophy. And *war* — massive, hostile, violent — *mostly male.*

I am confident that Nietzsche would say something very close to this. All human activities are expressions of the drive for power. War, or wars, are always exercises of power, always taking place within the context of the universal, eternal war of the sexes. War — the war we are trying to rethink — is about maintaining, or increasing dominance of the male over the female. The impulse, or will to power, in the clique of the inventors or instigators of war is, or becomes, contaminated with resentment — directed toward the powers, the abilities, of the woman and the mother/child relationship. The desire for revenge associated with resentment becomes lethal, venomous, viral.

Nietzsche, undoubtedly, would have contrasted the "real" war, his "truth," with the fabricated war — the images of war made up for the purpose of deceiving. Real war is dying, killing, burying, grieving, brutality, stupidity, futility. It is the devaluing of life, young life — immediate loss. And it deprives the mother of her "real" treasure — for a lifetime. Every lost life is a lost child. Every sacrificed life is a sacrificed child. The fabricated war, the false narrative, is about duty and glory. It makes killing and dying honorable. It justifies and celebrates death, not life. Violent war is invented and sustained by a web of deception.

In speaking about war, words matter. No one understood that better, or was better with words, than Nietzsche. And he did want others to listen to his words, to understand — but perhaps not right away. And to speak themselves, probably at some later date. Nietzsche understood that it matters *who* speaks, *what* is spoken about. Also, *when,* or *where,* or *why,* and *how.*

And most important, perhaps Nietzsche was the first modern philosopher to fully understand the *power* of words — what Merleau-Ponty would later refer to as "the power of signifying speech." Nietzsche believed that words were, or could be, more powerful than guns. Nietzsche covered his words with his "mask of contrariety." He posed riddles and puzzles — perhaps to attract his listeners to play his word games in probing for his serious intentions. Also, he assumed the persona of the god, Dionysos, whose priestesses and celebrants were the Maenads. Frequently portrayed as a the god of wine, the rose, the tree, all blossoming things, seasonal changes, Dionysos is often identified with the feminine, the passions, the unconscious, whose followers were believed to be women.

Increasingly, I have come to believe that Nietzsche/Dionysos was philosophizing about and for the future — or as a "prelude to a future philosophy." That indeed, his words were spoken mostly to woman, about woman, for woman — his future disciples. That he saw himself as becoming engaged in redeeming the creative power of speech of the female. His final words in *Ecce Homo* were, " — Have I been understood? — *Dionysos against the Crucified. . .*"

From *Ecce Homo,* listen again to a few of the opening words of the chapter entitled "Why I Am a Destiny":

> I KNOW my fate. One day there will be associated with my name the recollection of something frightful — of a crisis like no other before on earth, of the profoundest collision of conscience, of a decision evoked *against* everything that until then had been believed in, demanded, sanctified. I am not a man I am dynamite. . .I was the first to *discover* the truth, in that I was the first to sense — *smell* — the lie as lie. . . My genius is in my nostrils . . . I contradict as has never been contradicted and am nonetheless the opposite of a negative spirit. I am a *bringer of good tidings* such as there has never been, I know tasks from such a height that any conception of them has hitherto been lacking; only after me is it possible to hope again. With all that I am necessarily a man of fatality. For when truth steps into battle with the lie of millennia we shall have convulsions, an earthquake spasm, a transposition of valley and mountain such as has never been dreamed of. The concept

politics has then become completely absorbed into a war of spirits, all the power-structures of the old society have been blown into the air — they one and all reposed on the lie: there will be wars such as there have never yet been on earth. Only after me will there be *grand politics* on earth (*EH*, 126, 127).

Nietzsche's experiences clearly revealed to him that the traditionally enforced silence of the woman was being resisted, was coming to an end. That women were in the process of breaking this silence. The "convulsions," the "earthquake spasm" which Nietzsche sees in the future, suggest that he anticipates the time when females — *his disciples* — reclaim freedom of speech, the *power* of speech. Enter fully again into the "war of ideas" and the "*war*" between the sexes." New — *who* speaks. Chosen — *what* is spoken about. As Nietzsche was teaching, these women would speak, perhaps, of nature, life, sexuality, bodies, values, religion, power, the child, violence, truth, lies, war. Different experiences, different perspectives, different interpretations, different values, different visions of the future, different voices. *Mostly female.*

There are a few words that Nietzsche used in other contexts which he would have found appropriate in speaking of hostile, violent war — disgusting; poisonous; a violation of life; violation of the body, by "despisers of the body;" a curse; revenge against nature and against life.

At the beginning of this long, uneven, uncertain search for words about war, the first person to whom we appealed was Hesiod, "the poet of the roadside grass and the many-colored earth, and of men who live by the soil." Of whom it was written, "Echoes of ancient peasant wisdom, and of the mysteries of the earth, linger in his pages."

Nietzsche was a lover of poetry as well as being a poet himself. What follows is from another poet — of our own present era:

> *We shall not cease from*
> *exploration*
> *And the end of our exploring*
> *Will be to arrive where we*
> *started*
> *And know the place for the first*
> *time.*
>
> — *T. S. Eliot*

BIBLIOGRAPHY

Nietzsche's Writings

The Antichrist (PN), in *The Portable Nietzsche*, selected and translated, with an introduction, prefaces, and notes, by Walter Kaufmann (The Viking Press, Inc., 1968)

Beyond Good and Evil (BGE), translated and with an introduction by Marianne Cowan (Gateway Editions, Ltd., 1955)

Ecce Homo (EH), translated, with an introduction and notes, by R. J. Hollingdale, (Penguin Books, 1979)

The Gay Science (GS), Translated, with Commentary by Walter Kaufmann (Vintage Books Edition, Random House, Inc., 1974)

Human, All Too Human (HA), translated by Marion Faber, with Stephen Lehmann, introduction and notes by Marion Faber (The University of Nebraska Press, 1984)

The Portable Nietzsche (PN), selected and translated, with an introduction, prefaces, and notes, by Walter Kaufmann (The Viking Press, Inc., 1968)

Thus Spoke Zarathustra (Z), translated and with a Preface by Walter Kaufmann (The Viking Press, Inc., Compass Books Edition, 1966)

Twilight of the Idols (PN), in *The Portable Nietzsche*, Selected and Translated, with an Introduction, Prefaces, and Notes, by Walter Kaufmann (The Viking Press, Inc., 1968)

The Will to Power (WP), translated by Walter Kaufmann and R. J. Hollingdale, edited, with commentary, by Walter Kaufmann (Vintage Books Edition, 1968, Random House, Inc.)

Related Works

Approaches to Ethics (AE), *Representative Selections from Classical Times to the Present*, edited by W. T. Jones, Frederick Sontag, Morton O. Beckner, Robert J. Fogelin (McGraw-Hill, Inc., 1977)

Envy (E), A Theory of Social Behavior, by Helmut Schoeck, translated from the German by Michael Glenny and Betty Ross (Helmut Schoeck, A Helen and Kurt Wolff Book, Harcourt, Brace & World, Inc., New York, 1966)

Great Political Thinkers (GPT), *Plato to the Present*, by William Ebenstein (Fourth Edition, The Dryden Press, Inc., 1969)

Hesiod (H), translated by Richard Lattimore (The University of Michigan, 1959)

History of Ideas on Woman (IW), *A Source Book*, by Rosemary Agonito (Rosemary Agonito, Perigee Books, G. P. Putnam's Sons, New York, 1977)

Myth, Religion, & Mother Right (MR), *Selected Writings of J. J. Bachofen*, translated from the German by Ralph Manheim, with a preface by George Boas and an introduction by Joseph Campbell (Princeton University Press, 1967, First Princeton/Bollingen Paperback Printing, 1973)

Nietzsche (N), by R. J. Hollingdale (first published in 1973, Routledge & Kegan Paul Ltd.)

Nietzsche, Philosopher of the Perilous Perhaps, by Rebekah Peery (Algora Publishing, New York, 2008)

What is Called Thinking? (CT), by Martin Heidegger, English translation (Harper & Row, Publishers, Inc., First Harper Torchbook edition, 1972)

Leviathan (L), by Thomas Hobbes, edited by Michael Oakeshott, with an introduction by Richard S. Peters (Simon & Schuster, 1962, First Touchstone Edition 1997)

Leviathan (LE), by Thomas Hobbes, edited with an introduction by C. B. Macpherson (Penguin Books, 1981)

Sex and Power in History (SP), by Amaury de Riencourt (Dell Publishing Co., Inc., 1974)

INDEX